Long-Term Non-Operating Reliability of Electronic Products

Judy Pecht
Michael Pecht

Long-Term Non-Operating Reliability of Electronic Products

CRC Press
Boca Raton New York London Tokyo

Library of Congress Cataloging-in-Publication Data

Catalog record is available from the Library of Congress.

International Standard Book Number 0-8493-9621-2
Printed in the United States of America 1 2 3 4 5 6 7 8 9 0
Printed on acid-free paper

MOTIVATION

Although innumerable books and technical papers discuss the reliability of electronic components under operational conditions, there is a lack of information on electronics reliability under non-operating conditions. In general, electronics are developed and manufactured not to fail under operating conditions, such as current, voltage, and generated temperature. However, electronics are also subject to non-operating conditions (dormancy and storage), in which *stresses* such as humidity, ionic contaminants, temperature, radiation, shock, and vibration can occur over extended periods.

The need for high non-operating reliability of electronics is especially critical in lifesaving devices, such as air bags in automobiles, fire alarm systems, standby power sources, burglar alarms, earthquake warning systems, and radiation warning systems in nuclear power plants, which may lie dormant for a long time before being activated. Among defense applications, high non-operating reliability is important in missiles and munitions systems electronics, which often lie dormant for years before being deployed on very short notice.

WHAT THIS BOOK IS ABOUT AND HOW IT IS ORGANIZED

This book examines non-operating electronics reliability issues, outlining and discussing storage conditions, the stresses that can arise in these conditions, and the failure mechanisms that can cause a failure. **Storage** is the state in which the system is totally inactive and resides in a storage area. Often it is necessary for a stored device to be unpacked and connected in order for it to be tested. Other non-operating conditions include **dormancy** — the state in which the system is in its normal operational configuration and may be electrically connected but is not operating.

This book consists of nine chapters that can be read in any order. The order here walks the reader through the problems, examples, and solutions.

Electronics can experience a range of environments subsequent to manufacture and prior to disposal. Non-operating electronics do not necessarily experience benign environments. The potential environmental stresses on non-operating electronics can be natural, such as those due to climatic conditions, or can be induced by humans. Chapter 1 introduces the reader to storage environments and discusses the geographical impact on electronics.

Chapter 2 examines the two primary approaches used to address non-operating reliability of electronics. One approach uses models based on curve-fitting failure rate data obtained from field applications. The other is the physics-based approach which looks at the root cause of failure and provides time-to-failure models based on these physical phenomena.

Chapter 3 discusses the electrical stresses and failure mechanisms that affect electronics under non-operating conditions, including electrostatic discharge and contamination-induced parameter degradation.

Chapter 4 examines the origin and effects of corrosion. Corrosion phenomena are discussed and key failure mechanisms are explained.

Chapter 5 discusses the effects of radiation stresses on electronics and outlines the mechanical and electrical degradation of electronics as a result of exposure to radiation.

Chapter 6 addresses mechanical stresses in non-operating electronics, including temperature cycling and steady-state temperatures. The consequent failure mechanisms, such as fatigue failure, intermetallic formation, and cracking of packages are then presented.

Chapter 7 examines the long-term storage effects of specific electrical devices and assemblies. Parameters of concern along with available field data are also presented.

Since the time required to observe field failures under storage conditions is lengthy, accelerated testing procedures must often be used to obtain results. These tests should be carefully designed to prevent the occurrence of unusual failures under a particular accelerated stress. Chapter 8 presents accelerated tests utilized to assess non-operating conditions.

Chapter 9 outlines the various databases required to assess non-operating reliability and the framework of a reliability assessment methodology. The chapter also discusses the identification and integration of correlative parameters that make up a comprehensive model.

WHO THIS BOOK IS FOR

This book is important to several groups of readers: product reliability engineers faced with the task of providing long-term storage reliability of electronics; engineers who have encountered failures in non-operating equipment and would like to get to the root cause of the failures in order to weed them out; maintenance engineers and other personnel who need to maintain and service equipment to ensure high availability; readers who want to study the physics-of-failure approach to non-operating reliability; and students and researchers interested in studying different reliability aspects of electronics. It is also hoped that this book will stimulate further research in the field of non-operating reliability of electronics.

ACKNOWLEDGMENTS

Much of the research and preparation of source materials for the book was accomplished as a result of several SBIR projects. We thus want to give special thanks to the U.S. Army Armament Research, Development and Engineering Center (ARDEC) for supporting and encouraging the research that led to this book. In particular we thank Theodore J. Malgeri, Martin Melmed, Paul Granger, Gabriel Silberman, Joe Martinelli, and Cliffer Shook.

We are also indebted to a large number of people who supplied us with a wealth of technical information and other resource materials. These people include Dr. Donald Barker, Katherine Ross, Shrikar Bhagath, Jenny Chen, Richard Bauernschub, and Kevin Cluff.

Finally, we would like to express our appreciation to Iuliana Roshou Bordelon and Kim Jong who helped in the preparation of the manuscript.

JUDY AND MICHAEL PECHT

Table of Contents

To our parents, our brothers and sisters, and our children

Chapter 1

INTRODUCTION

Imagine the consequences of the failure of a fire alarm system installed in a residential building to go off in a fire emergency, or of an automobile air bag failing to inflate in a head-on collision. These are frightening examples of equipment failing to perform its required function after prolonged periods of non-operation (dormancy), often despite having passed quality and functionality checks at the time of installation. In both these cases, the problem can be attributed to either mechanical or electronic components that fail to activate.

1.1 THE NON-OPERATING ENVIRONMENT

In the voluminous literature on the reliability of microelectronics, much attention has been focused on failures due to operational stresses. However, relatively few studies have been devoted to addressing failures caused by non-operating stresses, often viewed as comparatively negligible. In particular, when microelectronics failure rates are presented, failures due to dormancy and storage effects are rarely investigated even though design engineering and logistics experience indicate non-operating conditions must be incorporated into product development and support [Ailles, 1989].

The need for non-operating reliability is especially important when electronic products are stored for months, or even years, and are then expected to perform satisfactorily without prior inspection.

Dormancy is defined as the state in which the equipment is in its normal operational configuration and connected, but not operating. For testing purposes, equipment in the dormant state may be cycled on and off. During dormancy, the electrical stresses normally experienced under operational conditions are usually eliminated or reduced.

Storage is defined as the state in which the system, subsystem, or component is totally inactivated and resides in a storage area. The product may have to be unpacked and connected to a power source to be tested.

Periods during which a product is non-operating, is considered a spare, or is moving to or from spare status, are not as innocuous as they seem. Indeed, parts can and often do degrade and fail while in dormancy/storage environments. In light of this, knowledge of the causes and effects of non-operating damage is needed for reliability assessment.

The reliability of a specific product will vary, based not only on the design and manufacturing process and materials, but also on its stress history due to repair, handling, operation, storage, and dormancy environments. Even storage or movement between storage sites and operation sites may introduce stresses that affect a product's reliability. A "dormant environment" can subject a product to stresses, including temperature, biological factors, humidity, and chemical reactions, which, depending on the design and manufacture of the product and its prior cumulative stresses, can significantly affect its reliability in future operations.

Harris [1980] has compiled and summarized well-known equipment classes (including commercial and consumer products) and typical values for the percentage of time spent in the dormant condition. This summary is given in Table 1.1, which shows that most industrial equipment has dormancy rates above 90%. Safety equipment has a dormancy rate of 98%.

1.2 NON-OPERATING ENVIRONMENTS

Products experience a range of environments subsequent to manufacture and prior to disposal. Non-operating parts do not necessarily experience benign environments, and dormancy is not limited to a single environment. On the other hand, not all dormancy environments merit consideration in determining reliability for each product; many are below the practical threshold for a specific part, either because the environment is sufficiently benign, or because the inherent characteristics of the product are not susceptible to the stresses associated with the environment.

The environmental stresses on non-operating products can be natural, such as those due to climatic conditions, or induced by humans. Natural environmental conditions include wind, precipitation, lightning strikes, earthquakes, temperature, humidity, vibrations, and radiation. Human activities include manufacturing, testing, pre-installation, operation, disposal, and mishandling of products by, for example, dropping them, brushing them against other objects, or tearing their protective covering - accidents that enable electrostatic discharge and contamination. Atmospheric gases, including ozone, sulfur dioxide, and particulate can appear in all non-operating environments and can reduce reliability.

Table 1.1 Typical Percentages of Calendar Time for Equipment in Dormancy [Harris, 1980].

Category	Percentage
Domestic appliances	
• Television sets	75%
• Kitchen electrical appliances	97%
Cars	
• Personal use	93%
• Taxis	38%
Professional equipment	
• Personal calculators	98%
• Small copying machines	> 75%
• Electronic test equipment	> 90%
Industrial equipment	
• Safety equipment	98%
• Standby power	> 90%
• Valves (most)	> 75%
• Air conditioning	50-80 %
• Built-in test equipment	99%

Figure 1.1 provides a matrix of possible stress factors and dormancy environments. The H, M, and L values in the matrix indicate high, medium, and low risk of exposure to stress-inducing activities in each dormancy environment. For each risk offsetting measures can be taken, such as providing protective coverings or containers. Mishandling or mismanagement can also introduce stress risks; for example, allowing a storage site to fall into disrepair or allowing water to enter the storage area can introduce a range of stress factors, such as humidity/moisture, biological agents, chemical reactants, and particulate contamination.

An explanation of these categories of non-operating environments and associated stress factors are given below.

1.2.1 Receipt Screening

Prior to being placed in inactive storage, parts are subjected to receipt screening, which can include removal of some or all of the protective containers and coverings. Electronic scanning devices can be used to assist with item identification. The receipt screening process is a transition point in the flow of material, and is therefore a location of such probable environmental stresses as thermal, biological, and humidity stresses, and acoustic noise. Human or mechanical handling at this stage can cause mechanical shock, physical impact, or particulate contamination.

Dormancy environments

Stress factors	Receipt screening	Inactive storage	Modification	Repair	Test	Movement	Awaiting transportation	Transportation	Inspection	Inventory action	Servicing in storage	Kitting	Troubleshooting
Thermal	M	L	M	M	M	H	H	H	L	L	L	M	M
Biological	L	L	M	M	M	M	M	M	L	L	L	M	L
Humidity/Moisture	M	M	M	M	M	M	H	H	L	L	L	L	L
Chemical reaction	M	L	M	M	M	L	M	H	M	L	L	L	M
Particulate contamination	M	L	M	M	H	M	M	M	M	L	H	M	H
Acceleration	L	L	L	L	L	M	L	H	L	L	L	L	L
Acoustic noise vibration	M	L	M	M	M	H	H	H	L	L	L	L	M
Mechanical shock	M	M	M	M	M	H	M	H	L	L	L	M	M
Radiation	L	L	L	L	L	L	M	H	L	L	L	L	L
Pressure	L	L	L	L	L	L	L	H	L	L	L	L	L
Physical impact	M	L	H	H	H	H	M	H	M	M	M	M	M
Ionizing radiation	L	L	L	L	L	L	L	L	L	L	L	L	L
Nuclear radiation	L	L	L	L	L	L	L	L	L	L	L	L	L
Electrical	L	L	M	M	M	L	L	L	L	L	L	L	M
Mechanical	M	L	M	M	M	L	L	L	L	L	L	L	H
Abrasion	L	L	L	L	L	L	H	H	L	L	L	L	L

Figure 1.1 Matrix of possible stress factors and dormancy environments

The receipt screening environment is relatively short because the receiving storage facility, as opposed to a general purpose warehouse, is dedicated to the storage of such products. A nominal estimated time frame for receipt screening is two days.

1.2.2 Storage

Products in storage can be kept in containers, drawers, shelves, or boxes. Storage sites can range from containers in an open field to environmentally controlled spaces shielded from external environments. Although the stored product itself can be static, the area in proximity to the product can be affected by nearby equipment associated with material storage or handling. Products can experience stress factors such as particulate contamination, shock, vibration, humidity, electromagnetic radiation from nearby emitters, or temperature extremes. An example of temperature extremes measured in several storage environments is given by the U.S. Army [1978]:

- Open storage: max., 165°F (74°C); min., -44°F (-42°C)
- Non-earth-covered bunkers: max., 116°F (47°C); min., -31°F (-35°C)
- Earth-covered bunkers: max., 103°F (39°C); min., 23°F (-5°C)

The Australian Ordnance Council [1983] found that solar radiation can also raise temperatures in excess of 30°C above the ambient. Pressure range extremes from 7.1 to 15.4 psi, and maximum absolute humidity at 13 grains/cu. ft. (relative humidity: 97% at 29.4°C), have been recorded [Sparling, 1967].

1.2.3 Modification

Modifications necessitate moving a product from its dormant status into a status in which changes are made, then to a transitional status in which the part is prepared for yet another status, such as transportation, storage, or operational installation. Products subjected to modification are often subjected to stresses associated with manufacturing or repair, as well as stresses from unpacking, repackaging, and the environment during assembly. The work areas for modifications can subject parts to a range of human-induced and environmental stresses, such as humidity, temperature, particulate contamination, mechanical shock, physical deformation, and electromagnetic radiation. Modification can also introduce replacement parts and materials with reliabilities different from those being replaced. Finally, post-modification testing can also introduce stresses.

1.2.4 Repair

Repair is similar to modification; however, parts entering repair status have usually failed or are part of a product or assembly which has failed. Testing for the cause of failure, replacing parts or materials, and retesting each part can introduce stresses. Human-induced and environmental stresses associated with modification can also apply to parts undergoing repair.

1.2.5 Test

Parts subjected to periodic testing, recertification, or requalification can be subjected to the same intrusive or non-intrusive stresses experienced by parts undergoing modification or repair. Tested parts, including spares, can be subjected to stresses comparable to or exceeding those associated with installed parts.

1.2.6 Movement

Parts in storage are often subjected to internal relocation within the storage facility. Facility modification, modernization, or maintenance; removal of a collocated part; and handling of the part incidental to operational installation or shipment can subject parts to a range of human-induced and environmental stresses such as humidity, temperature, mechanical shock, physical deformation, and electromagnetic radiation.

1.2.7 Awaiting Transportation

Parts in transition between two sites while awaiting transportation can be subjected to the most adverse stresses in the life cycle. Facilities at intermediate stops are more likely to have personnel inexperienced in the specific handling and environmental needs of the parts. Equipment, facilities, and environments may not be designed for the specific needs of individual parts. Radiation, thermal, moisture/humidity (rain), biological, particulate, mechanical impact, abrasion, and vibration stresses are highly probable at transshipment sites.

1.2.8 Transportation

Parts being transported by air, sea, rail, truck, or mail can be subjected to a broad range of extremely adverse stresses. Road bounce, rough terrain, and climatic conditions create stresses. Parts may also undergo thermal, biological, humidity, chemical reaction, particulate contamination, acceleration, acoustic noise vibration, mechanical shock, radiation, pressure, physical impact, ionizing radiation, electrical, mechanical, and abrasion stresses. Acceleration extremes have been measured in the transportation environment of up to

- 7 Gs at 300 Hz on trucks;
- 7 Gs at 1100 Hz on aircraft;
- 1 G at 300 Hz on railroad cars;
- 1 G at 70 Hz on ships.

and shock stress has been measured at 10 Gs on trucks and 300 Gs on railroad cars [U.S. Army, 1978]. Extreme temperature ranges associated with transportation and storage from -62°C (-80°F) to 71.1°C (160°F), and pressure ranges from 1.68 to 15.4, have been recorded [Sparling, 1967].

1.2.9 Inspection

Inspection of parts generally involves nonintrusively determining the integrity of external protective coverings or containers and the absence of physical damage. Unless mishandling occurs incident to inspection, the introduction of new stress factors is unlikely. If inspection involves intrusive measures that change the status of external containers or protective covers, biological, moisture, chemical, particulate, impact/deformation, electrical, or mechanical stresses can occur.

1.2.10 Inventory Action

Inventory action is similar to inspection, and can include inspection as well as parts counts. The purpose of inventory action is to determine the quantity of parts in inventory.

1.2.11 Servicing in Storage

Servicing products in storage can include represervation or repackaging, remarking external protective coverings, and replacing desiccants in protective containers. Stress conditions can be created by human handling, introducing materials like adhesives or particulates that can have secondary impact, disturbing factory installed protective coverings, and conducting intrusive inspections.

1.2.12 Kitting, Unkitting, and Restorage

Products placed in kits in anticipation of movement to other locations for forward support or mobilization are subjected to all the stresses associated with inactive storage, movement, transportation delay, and transportation, and may also be subjected to stresses associated with inspection, inventory action, and servicing.

1.2.13 Maintenance Troubleshooting

Spare parts used to calibrate or troubleshoot test equipment may be subject to stresses not otherwise found in dormant environments.

1.3 GEOGRAPHIC IMPACT

The life cycle environments of a product include extreme and varying climatic conditions due to geographic location. Exposure of products to climatic extremes and variations will impact reliability.

In 1979, the Army outlined climatic conditions according to climate type (AR 70-38). Climatic conditions include typical and extreme temperature and relative humidity readings, as well as the amount of precipitation, solar radiation, and wind. MIL-STD-210 is another source of climatic data for regions throughout the world and for suitable testing environments to determine design criteria. Pompei [1985] presented data on average hourly, daily, monthly, and yearly temperatures and relative humidities inside and outside ammunition depots throughout the United States.

Because of considerable differences in dormancy conditions with respect to temperature and humidity, the severity of exposure depends not only on geographical location, but also on conditions prevailing at the dormancy site. Factors introduced by geographic conditions, the actual building or shelter (tent, shed, or igloo), vibration, and shock must also be considered; for example, a quonset hut under the blazing sun is appreciably hotter inside than outside. The compounded effect of ambient temperature and solar radiation resulted in recorded diurnal temperatures of 75°C and -70°C at the external skin of U.S. Army missiles [U.S. Army, 1978]. Sometimes products are located in a canvas shelter in a tropical rain forest, with neither temperature nor humidity control; if the shelter has no air circulation, moisture will have a greater tendency to condense on surfaces, thus promoting corrosion. The range of recorded earth climate temperature extremes is 58°C to -80°C, an indication of the variation in temperature to which dormant parts can be subjected, and Resnick [1965] identified the extreme desert, arctic, and tropical climatic conditions to which microelectronics can be exposed.

1.3.1 Desert Environments

The desert environment is characterized by high temperatures (35 to 50°C), relative humidities of less than 30%, intense solar and ground radiation, minimal rainfall, large daily temperature changes of 20°C, and sandy dust. Direct exposure to solar radiation can raise the environmental temperature 30°C by magnifying the effects of the already high ambient temperature [Australian Ordnance Council, 1983]. When products are located in a shelter, even though the temperature never rises above the ambient, the combination of solar radiation and high air temperature can present problems. For example, in full solar radiation products can reach temperatures exceeding 70°C. Moreover, various parts of the products can attain different local temperatures in solar radiation, depending on the material's heat capacity, thermal conductivity, color, and access to shade.

Factors imposing loads in this environment include thermal stresses caused by solar radiation, ground radiation, and the ambient temperature. Strongly blowing sand and dust may lead to mechanical stresses. Such stresses may cause failures in several ways:

- Differential expansions and contractions may result in fatigue, distortion of assemblies, and rupturing of seals.
- Materials may deteriorate after prolonged exposure to ultraviolet rays and intense heat, which promote aging. This is particularly true of some elastomers that are degraded by ultraviolet light and ozone in the atmosphere [Jowett, 1973]. The effects of ultraviolet and infrared radiation are complex, because absorption is highly selective. Consequent chemical reactions and internal heating occur at different depths within the material in relation to the mechanical stresses

developed, which can lead to physical and chemical deterioration.

- High temperatures can lead to low viscosity, causing seal leakage and decomposition of organic materials.
- Sand and dust in the desert environment can erode surfaces, and powdery dust can penetrate into or between components that may appear sealed. Painted surfaces can be scoured clean of paint, and metals have been pitted and marked by sand and dust blown by high-velocity winds. Under such conditions, metal surfaces can be work-hardened during erosion. The low moisture content of the atmosphere can cause certain plastics to warp; some materials lose tensile strength, and materials incorporating organics, such as paper, can disintegrate.

1.3.2 Arctic Environments

The arctic and the desert similarly impose severe stresses on products, especially because at temperatures below -29°C snow takes on the character of sand, becoming particulate. Glacial dust, usually present in summer, closely resembles desert dust. These dust particles are alkaline, which can cause dust erosion and corrosive effects. The arctic environment is characterized by sub-zero temperatures, large daily temperature changes of 20°C in the spring and fall, and ice fog — all of which can contribute to a decrease in yield strength, ultimate strength, and ductility. Many polymeric materials, including plastics, become brittle to various degrees under such conditions, depending upon their molecular structure and composition. Low temperatures, particularly, cause a decrease in impact resistance and induce brittleness. The temperature coefficients of various components, such as capacitors, resistors, and inductance devices, can also change their performance.

Plastic seals at low temperatures can begin to leak due to shrinkage and cracking. Because water expands upon freezing, any moisture present can also have detrimental physical effects.

Generally, minimal corrosion of metals occurs at sub-zero temperatures, but in some parts of the arctic region, the presence of carpeted surfaces of thick moss, called muskeg, can turn water acidic, becoming a potential source of corrosion. Though the arctic waters are not considered harmful, muskeg waters with a high concentration of dissolved salts can be damaging. Condensation may arise from dense fogs that often cover polar coastal areas or when electronic devices are brought from the cold outside into a warm shelter, resulting in lead corrosion. The occurrence of "tin pest" is considerable at temperatures of 0°C, when a whiskery type of growth on tin-plated finishes can cause electrical breakdown in circuits by bridging conductors.

1.3.3 Tropical Environments

Tropical environments comprise high temperatures (31 to 41°C), high relative humidity (75 to 100%), daily temperature changes of up to 20°C, and frequent torrential downpours. As in the desert, solar radiation is a problem. But the parameters that differentiate the tropics from the desert and the arctic are high humidity and heavy rainfall, the basic requirements for corrosion and the presence of microorganisms. Products exposed to conditions of high humidity must be adequately protected. Cut edges of glass-fiber material must be sealed to prevent moisture absorption by capillary action. Flux and moisture entering the fractured wire termination seal of a plastic-encapsulated capacitor have also led to breakdown [Jowett, 1973].

Tropical soils are known for the presence of a large variety of microorganisms, bacteria, fungi, insects, and destructive termites, which can degrade many organic and inorganic materials. Fungi are a group of microorganisms that attack and digest all types of organic materials; conditions during dormancy in a tropical environment are particularly conducive for their rapid growth. Another, perhaps surprising, source that helps mold growth is tobacco smoke; tests have shown that mold grows faster on the assembled surfaces of materials, such as plastics, after direct contact with tobacco smoke. Degradation of physical properties by bacterial action involves not only naturally occurring fibers, but also synthetic materials. The admixtures found in the polymers (fillers, plasticizers) are usually the first to deteriorate under this influence.

In tropical environments metals corrode more rapidly, and electrolytic action between dissimilar metals is significantly accelerated. The surface and volume resistivity of insulating materials is lowered due to moisture absorption, an increase in which can cause swelling of materials, leading to electrical and mechanical breakdown. The presence of fungi and mold can be destructive due to the metabolic process associated with growth, which can cause etching on material surfaces. Also, fungal growth forms a low-resistance path [Jowett, 1973].

In some areas, products are covered with condensation at least once a day, as the temperature drops and the relative humidity rises each night. In marine areas, salt spray combined with high relative humidity can accelerate corrosion by acting as a catalyst. A significant amount of salt is carried by the winds from the ocean; however, corrosion depends more on the occurrence of salt spray than on the salt content of the seawater. Aluminum and aluminum alloys that resist humidity corrosion are prone to pitting corrosion due to salt and coral.

Table 1.2 Estimated Extreme Environmental Parameters for Army Tactical Missiles [Livesay, 1978].

Parameters	Estimated values
Maximum temperature	+75°C [a]
Minimum temperature	- 50°C [a]
Temperature cycling	ΔT ≈ 70°C [a]
Moisture	up to 100% humidity
Thermal shock	small
Mechanical shock and vibration due to transportation and handling	≈ 10g
Bacteria and fungus	heavy exposure
Nuclear radiation	not applicable
Electromagnetic fields	not applicable

[a] These values vary in different parts of the system.

1.4 DORMANCY ENVIRONMENT PARAMETERS AND VALUES

Some anticipated extremes in environmental parameters for Army Tactical Missiles have been compiled by Livesay [1978] and are given in Table 1.2. He warns, however, that it could be inappropriate to make drastic material changes in the design stage based on this information, because these extremes might not all occur simultaneously, and individual extremes may have a low probability of occurrence.

A valuable source of data representing various storage and transportation conditions at locations around the world is the study conducted by H.C. Schafer at the Naval Weapons Center at China Lake, California. Kurotori [1972], Riordan [1974], Blackford [1972], and Hirschberger [1975] also have compiled a significant amount of data on weather patterns and environmental extremes.

1.5 FAILURE MECHANISMS

In the concluding text of this chapter, stress conditions and failure mechanisms unaffected by long-term storage and dormancy are briefly mentioned, with an explanation of why they are excluded. In subsequent chapters, the focus will thus be on the key failure mechanisms that influence the reliability of electronic products in a non-operating environment.

1.5.1 Electrical Failure Mechanisms

In general, electrical failure mechanisms do not occur in non-operating conditions, but occur during operation or testing, when an external voltage or current is applied. Electrical failures mechanisms which do not arise in non-operating conditions include:

- Electrical overstress, which results from higher-than-rated voltage or current induction of a hot-spot temperature;
- Time-dependent dielectric breakdown; i.e., the formation of low-resistance dielectric paths through localized defects in dielectrics;
- Secondary breakdown, which results from an increase in current density and temperature, or a drop in voltage in power transistors;
- Electromigration, which results from high current density in metallization tracks, producing a continuous impact on metal grains in the tracks, causing the metal to pile up in the direction of current flow, and producing upstream voids;
- Hot electrons, which are charge carriers that acquire energy from very strong electric fields, cross the potential barrier at the silicon-silicon oxide interface, and cause threshold voltage shifts;
- Hillock formation, which results from electromigration; and
- Metallization migration, which occurs when there is a certain level of current density at a dendrite tip, sufficient liquid medium, and an applied voltage that exceeds the sum of anodic and cathodic potentials in equilibrium with the electrolyte.

Several failure mechanisms occur only from extremely high temperatures (175 to 400°C), uncommon to most non-operating environments. These include:

- Contact spiking, which is the penetration of metal into the semiconductor in contact regions;
- Surface charge spreading, which is the lateral spreading of ionic charge from biased metal conductors along the oxide layer or through surface moisture; and
- Slow trapping, whereby electrons are trapped when they cross the potential barrier of the silicon-silicon oxide interface, causing a shift in the threshold voltage.

Electrical failure mechanisms for non-operating conditions are discussed in Chapter 3. Electrical failure mechanisms influenced by non-operating conditions include electrostatic discharge caused by random electrostatic pulses from the outer environment and ionic contamination.

1.5.2 Corrosion Failure Mechanisms

Corrosion — including galvanic, pitting, crevice, and stress corrosion — can occur during dormancy without an applied voltage. While numerous factors promote corrosion, the degree of hermeticity which a product can achieve will determine the amount of moisture and contaminants that enter the product. In addition, microorganisms growing on and inside the product can produce acids corrosive to metals, glass, and organic material. However, electrolytic corrosion, which requires an externally applied voltage across an electrolyte between adjacent bond pads and metallization tracks, will not occur in unoperated dormant products.

Three failure mechanisms must not be mistaken for corrosion: intermetallic formation, whisker growth, and damage which occurs from improper manufacturing methods, such as abusive ultrasonic cleaning. Under a scanning electron microscope, intermetallic formations appear to be corrosion by-products, but they are not. Whisker growth, promoted by high applied or residual stresses, occurs as a result of recrystallization at room temperature. Although humidity encourages whisker growth and is similar to corrosion, whisker growth should not be confused with corrosion. Finally, as an example of problems with manufacturing, Sandia National Laboratories reports that chip carriers subjected to ultrasonic cleaning of electronic components and assemblies, and taken out of dormancy, began to shed their leads. Although corrosion products were found, fatigue striations indicated ultrasonic cleaning as the true source of damage [Cieslak, 1987]. Corrosion failure mechanisms for non-operating conditions are discussed in Chapter 4.

1.5.3 Radiation Failure Mechanisms

Radiation can cause mechanical and electrical degradation. Mechanical degradation consists of point defects that lead to material embrittlement and alter the properties of the materials. Electrical degradation which can only occur during operation, causes a temporary change in the logic state of the bit; however, the radioactive isotopes, mainly alpha particles, that cause electrical degradation can build up during long-term storage and dormancy and precipitate failures during subsequent operation. Therefore, the accumulation of alpha particles becomes a significant concern. Radiation failure mechanisms for non-operating conditions are discussed in Chapter 5.

1.5.4 Mechanical Failure Mechanisms

Mechanical failures under non-operating conditions result primarily from temperature cycling, vibrations, and shock. These give rise to expansion mismatches between the various parts, which in turn cause fatigue and fracture. Reversion or depolymerization, the breaking of polymeric bonds in printed wiring board materials, die attaches, conductive adhesives, and the encapsulant of plastic-encapsulated devices should not be used if they degrade at temperatures anywhere near the highest temperature recorded in most non-operating environments. Mechanical failure mechanisms for non-operating conditions are discussed in Chapter 6.

Chapter 2

NON-OPERATING RELIABILITY ASSESSMENT

This chapter discusses approaches to assessing the reliability of electronic products under non-operating conditions. The first four approaches use "phenomenological" models based on failure rate data obtained from field data, while the last approach is a physics-based approach that looks at the root cause of failure of each product and provides assessments and solutions based on physical phenomena.

2.1 THE RADC-TR-85-91 METHOD

RADC-TR-85-91, titled "Impact of Non-operating Products on Equipment Reliability", was published in May 1985 as a means to calculate the total storage failure rate for some piece of equipment and predict if the equipment meets the storage reliability goal. Regression analysis and other statistical techniques were applied to field data of non-operating equipment and empirical failure rate models were developed for each class. In general, each model has a base non-operating storage failure rate, modified by factors of temperature, quality, and on-off cycling. The document defines 26 different environmental factors; the models are modified to account for factors specific to each kind of component. The general storage failure rate model used in this method looks like:

$$\lambda_p = \lambda_{Nb}\pi_{NT}\pi_{NE}\pi_{NQ}\pi_{cyc} \qquad 2.1$$

where λ_p is the product non-operating failure rate, λ_{Nb} is the base non-operating failure rate, π_{NT} is the non-operating temperature factor, π_{NE} is a non-operating environmental factor, π_{NQ} is a non-operating quality factor, and π_{cyc} is the equipment power on-off cycling factor.

The cycling factor is added to account for dormancy conditions. Since equipment in a dormant state is in a normal operating configuration and connected, though not operating, it may be cycled on or off for testing purposes, which could cause more damage than lengthy storage times [Rooney, 1989].

The most severe limitation of this method is the assumption that storage failures of all products are exponential, with a constant failure rate. This assumption fails with products with a finite shelf life, like aluminum electrolytic capacitors, whose shelf life is a function of the rate of expenditure of the electrolyte — again, a function of the local ambient storage temperature and the purity of the aluminum used in manufacturing the capacitor [Rooney, 1989]. Batteries, various electronics which use adhesives, and products which fail due to cumulative stresses, also fall into this category.

Gross inaccuracies are another severe limitation, partly because the document is so out-of-date and partly because key manufacturing information was never incorporated into the models.

Other limitations are associated with the practical use of this method. For example, it does not account for specific storage conditions, such as those for sonobuoys, which are normally stored in "overpacks" that provide both a moisture barrier and a barrier to contaminant gases [Rooney, 1989]. This environment is somewhat less harsh than "ground, fixed" and more harsh than "ground, benign", yet one of these has to be chosen in order to make failure rate calculations. Environmental factors alone, then, do not fully describe the situation.

2.2 MIL-HDBK-217 "ZERO ELECTRICAL STRESS" APPROACH

MIL-HDBK-217 [1991] provides failure rate models based on curve-fitting empirical data obtained from operational tests conducted on electronic components. The models have a base failure rate modified by a product of environmental, electrical stress, quality, and other factors. A model for non-operating conditions can be obtained by eliminating from the model all factors dependent on operation, such as electrical stresses [Harris, 1980]. The remaining equation has a similar form and the same problems as does the RADC-TR-85-91 method. In addition, while the approach is used by some military contractors, this is not a valid approach to predicting failure rates in dormant/storage conditions, because empirical relations cannot be extrapolated beyond the range in which the data were gathered. As a result, most predicted failure rates are extremely inaccurate, and because they are neither conservative nor optimistic, cannot be used for any trade-offs [Cushing, 1993, 1994].

2.3 THE "K" FACTOR APPROACH

This approach assumes that there is a direct relationship between the operating and non-operating reliabilities of electronics, and that the two are directly proportional. Hence, a multiplicative factor "K" is applied to the operating failure rate in order to obtain the non-operating failure rate. A crude rule of thumb, used before any systematic study of storage reliability was conducted, was that the ratio between operating and non-operating failure rates was about 10 to 1. However, after the completion of studies on storage reliability, the 10 to 1 ratio was considered very pessimistic. A more realistic ratio of 30 to 1 or even 60 to 1 was predicted for most electronic components [Harris, 1980]. For electronic assemblies, ratios as high as 80 to 1 were considered typical.

This approach is at best a very crude approximation, because the stresses that affect the operating and non-operating reliability of a component are definitely not the same. Hence, it is erroneous to assume that the two rates are proportional.

2.4 THE MARTIN-MARIETTA TEST PROGRAM

In May 1965, under an Air Force award to gain a better understanding of non-operating failure rates in electronic equipment [Cottrell, 1967], researchers at Martin-Marietta collected non-operating data in excess of 760 billion part-hours with a maximum storage time of seven years. Eight different hardware groups were selected, including printed circuit boards, chassis assemblies, ultrasonically cleaned PCBs, micromodules, test console parts, and microcircuits. All of these parts were tested prior to storage, then again after spending time in a storage environment. Not all parameters were measured prior to placing the hardware in storage, and hence the data represented only the measurements that could be made within the limitations imposed by the earlier test programs [Cottrell, 1967]. In the case of microcircuits, for example, they conducted a drift analysis on a few parameters such as output level-off voltage and input turn-off current and they concluded that some parameters can vary erratically during storage and still remain in specification, while others tended to drift consistently in a particular direction.

Additionally, the researchers gathered data on both accelerated and non-accelerated aging tests conducted on microcircuits from both vendor and nonvendor sources, and analyzed the accelerated data based on the Arrhenius theory. They concluded that both vendor and nonvendor microcircuits showed a decreasing linear failure rate with the reciprocal of the absolute temperature. They then tabulated non-operating failure rates for electronics under two categories: catastrophic and drift failures. The tabulations indicated that for the ratios for electronic systems employing military standard parts, catastrophic non-operating failure rates of 4.7 fits/part were calculated, while for analog electronic systems utilizing military standard parts, the ratio of drift to catastrophic non-operating failures was calculated to be approximately 1 to 25.

A typical example of a model developed as a result of this study is shown in Figure 2.1, for both test and nontest conditions [Cottrell, 1967]. In the reliability model the term a is 90 % (portion of failures detected by periodic test equipment), t_1 is 43,800 hours or 5 years (total non-operating time prior to launch), t_2 is 41,610 hours or 4 years, 9 months (non-operating time up to last periodic test before launch), t_3 is 2190 hours or 3 months (non-operating time between last periodic test and launch), t_4 is 0.0166 hour or 1 minute (flight time), μ_1 is 8,147 fits (sum of non-operating failure rates for 1539 missile electronic parts), μ_2 is 634,244 fits (estimated operational failure rate values are normally obtained from sources such as MIL-HDBK-217), and k is 1000 (factor for flight environment).

Summing up the no-test terms,

$$\lambda t = \mu_1 t_1 + k\mu_2 t_4 = 0.357 + 0.011 = 0.368 R = e^{-\lambda t} = 0.692 \tag{2.2}$$

For the three-month test condition,

$$\lambda t = (1-a)\mu_1 t_2 + \mu_1 t_3 + k\mu_2 t_4 = 0.034 + 0.018 + 0.011 = 0.063 R = e^{-\lambda t} = 0.939 \tag{2.3}$$

Again, the problems associated with this approach are the same as those described with the first three approaches.

2.5 THE PHYSICS-BASED APPROACH

Cherkasky [1970] has stated in a nutshell the main problem with all the methods mentioned so far:

> "Nearly all existing non-operating failure rate data lack a common base line or a set of fixed ground rules. For example, if one were to choose at random three independent storage reports containing storage data, it would be doubtful that any have defined what constitutes a storage induced failure, let alone delineate all pertinent ground rules or test conditions. Data summarized from many sources are therefore of dubious value for meaningful system predictions."

The final objective of any reliability prediction method is to prevent any design, manufacturing, and operational failures associated with the product. Failure rate predictions of electronics based on empirical modeling of field failure data, using curve-fitting techniques, do not take

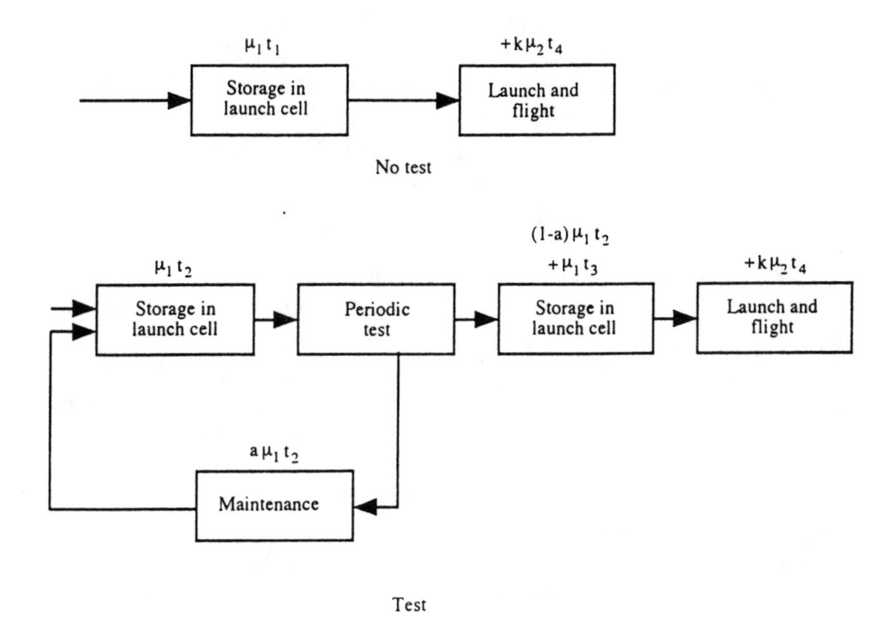

Figure 2.1 The Martin-Marietta model [Cottrell, 1967].

into consideration crucial failure details such as failure mechanisms, failure modes, load/environment history, package materials, and package geometry. They tend to group similar products under a common category, regardless of specific differences in their geometry, materials, and architecture. This methodology also does not address design and usage parameters that have a great influence on reliability. Thus, the very purpose of conducting a reliability prediction — that is, to prevent the recurrence of the failure under similar conditions — is defeated. Since no failure mechanism and stress information is available to support these predictions, it is extremely difficult, if not impossible, to pinpoint the actual cause of failure and to provide a remedy.

Physics of failure is an approach to design, reliability assessment, testing, screening, and stress margins that uses knowledge of root-cause failure processes to prevent product failures through robust design and manufacturing practices. Pecht [1994a,b], Cushing [1993, 1994], and Lall

[1993] have outlined the approach for conducting a physics-of-failure study, including the following steps.

2.5.1 Define Realistic System Requirements

Every product must operate through a range of application conditions for a specified length of time. Requirements may be determined by mission profile, number of missions or required storage and operating life, performance expectations, size, weight, or cost. The manufacturer and the customer must jointly define system requirement in the light of both the customer's needs and the manufacturer's capabilities to meet those needs.

2.5.2 Define the System Usage Environment

The manufacturer and the customer must jointly specify all relevant operating, shipping, and storage conditions, and assess available tradeoffs. The traditional use of standard environmental categories should be replaced by measured data for temperature, temperature changes, operable duty cycle, humidity, vibration, applied voltage, and any other key electrical, thermal, radiation, and mechanical items.

2.5.3 Identify Potential Failure Sites and Failure Mechanisms

Potential failure modes, sites, and mechanisms, as well as potential architectural and stress interactions, must be identified early in the design process. Ishikawa diagrams and failure mode effects and criticality analysis (FMECA) can be used for this purpose. Pareto diagrams can be used to assess the relative importance of each type of failure and to determine a hierarchy of design features. Once expected failures are identified, appropriate measures must be implemented to reduce, eliminate, or accommodate them.

2.5.4 Characterize the Materials and the Manufacturing and Assembly Processes

All materials must be characterized and their key characteristics controlled. These include types and levels of defects, as well as expected variations in properties and dimensions of materials and in manufacturing and assembly processes. These variations can significantly impact product performance over time.

2.5.5 Design Reliable Products Within the Capabilities of the Materials and Manufacturing Processes Used

The design must be evaluated and optimized for manufacturability, quality, reliability, and cost effectiveness before production begins. It is unrealistic and potentially dangerous to assume structures are defect-free, because materials often have naturally occurring defects, and manufacturing processes can induce additional flaws. These concerns can be addressed concurrently, using experimental step-stress and other accelerated life-testing methods.

2.5.6 Qualify the Manufacturing and Assembly Processes

All manufacturing and assembly processes must be optimized and capable of producing the product. Key process characteristics must be identified, measured, and optimized.

2.5.7 Control the Manufacturing and Assembly Processes

The manufacturing process must be monitored and controlled. Defect-specific screening, based on the results of the characterization, design, and qualification phases can detect and screen out manufacturing defects, as appropriate.

2.5.8 Manage the Life Cycle of the Product

Closed-loop procedures must be used to collect data from tests performed in design, manufacturing, accelerated life testing, and field operation, and to continuously assess and improve the quality, reliability, and cost-effectiveness of the product.

This approach

- Proactively incorporates reliability into the design process by establishing a scientific basis for evaluating new materials, structures, and electronics technologies;
- Provides information to plan tests and screens, and to determine electrical and thermomechanical stress margins;
- Uses generic failure modes that are as effective for new materials and structures as they are for existing designs;
- Encourages innovative, cost-effective design through the use of realistic reliability assessment.

Chapter 3

ELECTRICAL FAILURE MECHANISMS

Although perhaps not readily apparent, non-operating electronics can be subject to significant electrical failure mechanisms. While electrostatic discharge (ESD) is the most common in storage/dormancy environments, electrical failure can also be caused from static charging of dust particles and organic vapors. This chapter discusses the nature of electrostatic discharge, the environmental causes, the physical effects, and reviews common ESD testing models. A brief overview of contamination-induced parameter degradation is also presented.

3.1 ELECTROSTATIC DISCHARGE

Large electrostatic charges can be developed when two different materials come into frictional contact and are then separated. Electrostatic pulses can arise from contact with air, skin, glass, or charge-carrying particles. Resulting voltages are a strong function of the relative humidity and can be as high as 35 kV [Moss, 1982]. The grounding of a charged device resulting in the redistribution of the charge is known as electrostatic discharge (ESD). Catastrophic or latent damage to the integrated circuit can be caused by the charge stored on a human body being discharged through the device to a ground, or similarly by the charge on the device grounded through one of its terminals. Devices sliding inside nonconductive handling containers or across benches can also develop large static charges sufficient to degrade CMOS device parameters or to break down gate insulators [Pancholy, 1977].

Many materials used in the fabrication and packaging of integrated circuits and semiconductor hybrids are susceptible to electrostatic charging. In the triboelectric series, air and human skin are the most positive charge-producing materials while silicon, Teflon[R], and silicone rubber are among the most negative. An integrated device can be considered equivalent to a circuit with multiple paths to the ground. When one pin is grounded, potentials sufficient to cause dielectric breakdowns, junction shorts, or cracks between isolated regions may be discharged through the product.

ESD failure types can be generally categorized as thermally or field-induced. Examples of thermally induced modes are filaments in the poly gate, contact metal burnout, melted filaments in the drain junction, or fused metallization. Field-induced failure types include latent hot carrier damage and dielectric breakdown. ESD damage to semiconductor devices is typically isolated to small internal locations, as opposed to EOS damage which covers large surface areas. The failures can also be latent, as when an oxide or junction degrades gradually until it fails catastrophically at some later time.

Electrostatic discharge is also a frequent cause of failure in linear bipolar integrated circuits, where only a minor shift in product characteristics can cause circuit malfunction. In fact, one-third of the bipolar integrated circuit failures detected in telecommunications systems during assembly are due to ESD-damaged products. Two mechanisms cause most ESD-induced IC failures. One is dielectric breakdown, the common failure mode in MOS ICs; the second is silicon melting, caused by localized overheating in the depletion region of a PN junction.

Dielectric breakdown of the gate insulator, leading to catastrophic failure, is a major reliability factor in metal-oxide semiconductor devices (MOS) and integrated circuits. Extraneous high voltages can cause dielectric breakdown due to the high impedance of the gate insulator and the low leakage characteristics of the thermally grown silicon dioxide. Discharge of the self-capacitance of IC packages or circuit assemblies can also destroy ICs [Minear, 1977].

Electrostatic pulses can also cause cracking of the device oxide or melting of small amounts of the device, and create minute explosions on the product surface, resulting in voids, cratering, and subsequent shorts or open circuits. Increased temperature causes a significant reduction in the electrostatic discharge resistance of the component [Kuo, 1983; Hart, 1980].

Low-voltage electrostatic pulses can damage gate oxides to varying degrees in NMOS without actually causing complete gate oxide breakdown [Amerasekera, 1986, 1987]. Damage shows up in the form of a reduction in saturation drain current and gate voltage. Scherier [1978] has shown that MOSFETs are most susceptible to damage from electrostatic discharge. Total breakdown of n-channel MOSFETs occurs around 100 V, while the TTL NAND gate device shows degradation around 1.0 kV. Latent damage can also occur when dormant microcracks are aggravated.

The electrostatic properties of dust can also damage electronic devices. One aspect of static in moving dust particles is the wide spectrum of phenomena that can occur, from microscale electrostatic copying to surface sticking. Dust particles become charged depending on ambient humidity, particle-size distribution, and the velocity with which the dust particles move or impinge on surfaces. Surfaces quickly become coated with fine particles in the presence of moving powders, which generally reduce the charging or cause polarity changes, since impacts will be between different-sized particles rather than between the surface and the particles. The static electrification binding forces of these particles can sometimes exceed gravitational forces. Packed dust particles may alter resistivities and hysteresis may occur. In dormant environments, therefore, care should be taken to prevent the intrusion of dust particles into packages [Jowett, 1976].

Static charging also occurs due to indirect causes, such as piezo and pyro-electrification. Some crystalline structures will become charged due to the mechanical stresses or molecular reorientation caused by applied pressure or heat. When a material is deformed or fractured, charge generation can also result from reorientation of polar molecules or ionic displacement. Crystals exposed to pressure can also exhibit an electrostatic potential when exposed to temperature gradients [McAteer, 1989]. This situation can arise in dormancy when the products are partially exposed to either heat or cold, or when winds circulate the air in the non-operating site, creating a temperature differential.

Temperature differentials in the dormancy environment can cause condensation. Electronic products are as prone to this phenomenon as is a glass of cold water in a warm room; yet the results are often much more detrimental for the electronic components. For example, when electronic products are brought into a warm shelter from the cold, a metal- in-solution can result, a problem that occurs when a metal dissolves in liquid with a high dielectric constant, such as water. The metal tends to go into solution in the form of positive ions or complex negative ions, depending on the dissolved substances present in the liquid. This solution process continues until the metal acquires a fixed potential. The problem becomes complex when two metals go into solution and acquire different potentials, creating a galvanic cell. This can happen during storage/dormancy, but may not be identified until an electric contact is made through a circuit when the electronic product is taken out of storage/dormancy and made operational. Consequently, a current results that allows more ions to be released from each metal in solution until numerous degradation mechanisms take place, eventually leading to the failure of the electronic product.

Electronic products are vulnerable to lightning attacks that lead to electrostatic discharge. Lightning induces large voltage and current surges in the electronic equipment. It is impossible to protect equipment connected directly to antenna systems or long exposed lines; if such equipment is poorly grounded, the damage may extend into the system and destroy related equipment. Ground connectors are rarely sufficient to accommodate the kind of massive grounding straps required [Arsenault, 1980].

Both internal and external ESD protection methods are important. Typically, present-day products are protected against ESD by structures that shunt the ESD event to a ground bus in order to limit the current entering the critical junction [Duvvury, 1983, 1993]. Good design practices require a safe ESD path between each pin and every other pin in the device. Damage voltages for protected products are in the range of 3000 to 9000 V, depending on the protection, compared with 100 V for unprotected products. For this reason, ESD is not usually a dominant failure mechanism in properly protected products. In advanced CMOS technologies, adequate protection becomes even more critical with higher current densities. ESD stress currents become more concentrated in those MOS technologies with shallow LDD junctions and silicided diffusions [Chen, 1986; Duvvury, 1986]. Although on-chip protection can reduce product performance and consume die area, it is a necessary price to avoid ESD damage.

External protection methods generally involve controlling static in the environment and storing the devices and assemblies in a Faraday shield. Complete assemblies must be placed in antistatic bags if they are to be stored or transported. However, antistatic bags are more successful in preventing static charging than in providing static protection. Static-protective containers must provide protection against triboelectric generation, electrostatic fields, and direct discharge from contact with charged people or objects. Triboelectric charge generation is a friction process dependent on humidity, materials, surface characteristics, geometry, and so on. Increasing the material's lubricity is one way of reducing friction. Antistatic bags, when used in dormancy environments, should be inspected for creases (especially on the bottom), because creases decrease the effectiveness of the conducting layer.

Microelectronics protected in antistatic bags for extended periods face several problems. First, low humidities, chemical contamination, and outgassing cause a loss in antistaticity. Antistatic materials are impregnated with antistats that migrate constantly to the surface, forming a sweat layer that increases the material's lubricity. These hygroscopic materials attract moisture from the surrounding air, so their effectiveness decreases when the relative humidity is low. Most antistatic plastics lose effectiveness from contact with paper products or exposure to air. Periodic tests may be required to check their effectiveness.

The accumulation of dirt, oils, and silicone during dormancy also has an adverse effect on the effectiveness of hygroscopic antistats. Since hydrocarbons such as alcohols and ketones are used for cleaning them, the use of damaged bags would lead to a new problem involving organic vapors.

Antistats used in some hygroscopic materials can track onto items and act as a foreign substance that can react adversely with other materials

[Edwards, 1982]. Test results also show that the useful life of antistatic bags is limited; certain bags showed aging after a period of one year [Head, 1982]. Antistatic polyethylene material can also cause contamination problems: "The solder coating on part leads was depleted by reaction with a contaminant. Studies showed that lead di(n-octanoate) was the cause whose source was traced to n-octanoic acid contained in the antistatic agent used in polyethylene" [GIDEP Alert Number E9-A-86-02 ESD 431]. In addition, a commercial brand of MIL-B-81705 type-II film contained organic acid and caused corrosion of solder-coated product leads on circuitry. Even acid-free antistats can stress-crack polycarbonate, weaken adhesive bonds, and discolor epoxy paint [Koyler, 1982].

Three common models describe ESD pulses during handling: the human body model (HBM), machine model (MM), and charge device model (CDM) [MIL-STD-883C; Klinger, 1990; Renninger, 1989; Rozozendaal, 1990]. The human body model simulates the charge transfer pulse that a human body would inflict by touching the electronic device. The CDM models the effect of charging a device, which can happen during the assembly or shipping, and then discharging it to ground. The machine model simulates the charge transfer to a device in assembly, bonding, or test operations. During storage, ESD damage from any of these stress conditions is possible. Properly designed protection circuitry will allow the device to survive all of these conditions.

The HBM's model parameters for the human body are a source voltage of from 1 to 4 kV, 100 pF capacitance, and a resistance of 1500 Ω. The human body model has been standardized per MIL-STD-883C. In the charge device model, the electronic device is charged to a level representative of device testing. It is subsequently discharged to ground in about 5 ns. The most widely used specification is from AT&T microelectronics [Klinger, 1990; Renninger 1989]. The MM-ESD model does not have the capacitance or resistance of the human body; thus, although the voltage is lower (about 200 V) the currents are considerably higher, 1 to 10 A for the first peak values. Japan and Phillips Research Laboratories produced the two common MM specifications [Rozozendaal, 1990].

3.2 CONTAMINATION-INDUCED PARAMETER DEGRADATION

Contaminants that enter an electronic product can cause ionic contamination, which in turn can cause reversible degradation phenomena, such as threshold voltage shift and gain reduction [Schnabble, 1988]. A survey of published data shows that about 30% of microelectronic MOS product failures are a result of ionic contamination [Brambilla, 1981, Johnson, 1976]. Ionic contamination is caused by mobile ions in semiconductor products. Contamination can arise during packaging and interconnect processing, assembly, testing, screening, and operation. Sodium, chloride, and potassium ions are the most common contaminants; the most dangerous is sodium, which can cause failures in extremely small quantities ($10^{11}/cm^2$). The amount and distribution of alkali ions (Na^+, Li^+, K^+) in or near FET-gate dielectric regions influence the product threshold voltage by superimposing ionic charges on externally applied FET product voltage. An additional positive charge at the silicon/silicon oxide interface induces extra negative voltage in the n-channel, resulting in a decrease in the threshold voltage of the product.

Surface-charge spreading is largely observed in MOS and memory products and involves the lateral spreading of ionic charge from biased metal conductors along the oxide layer or through moisture on the product surface [Edwards, 1982; Blanks, 1980; Stojadinovic, 1983]. An inversion layer outside the active region of the transistor is formed by the charge, creating a conduction path between the two diffused regions or extending the p-n junction through a high leakage region; this results in leakage currents between neighboring conductors. The rate of charge spread increases with temperature. Surface-charge spreading failure, a wear-out mechanism, is usually observed at temperatures around 150°C and 250°C [Lycoudes, 1980].

In general, a high-temperature storage bake and exposure to high temperature during burn-in screens out ionic contamination failures [Hemmert, 1981; Bell, 1980]. In particular, a uniformly distributed contaminant will greatly reduce its effectiveness in altering the threshold voltage.

Chapter 4

CORROSION FAILURE MECHANISMS

Electronic products have become more vulnerable to corrosion due to dense packaging and high-impedance circuitry. Encapsulating larger chips in smaller packages reduces design allowances for thermomechanical stability and protection from the environment; narrow electrical pathways and connections can tolerate very little corrosion before circuit function is impaired [Kinsman, 1987]. Corrosion becomes more acute as product metallization tracks become narrower, the separation between metallization tracks becomes closer, and bond pads become smaller, because the smaller dimensions (thin-films can be as thin as 200 nm) lead to faster rates of corrosion [Devanay, 1989]. Gradually decreasing lead diameters mean less metal is available to withstand corrosive attack before lead failure occurs [Berry, 1987].

Corrosion, typically defined as the chemical or electrochemical reaction of a metal with the surrounding environment, disables electrical circuits and degrades the magneto-optic properties of rare-earth/iron-alloy amorphous thin-films by air oxidation [Frankenthal, 1987]. The continuation and rate of the corrosion process depend on the nature of the corrosion product. Conditions that accelerate corrosion include relative humidity, high temperatures, high contaminant concentrations, and the presence of dirt or dust, which can hold more moisture on the surface of the metal. In addition, duty cycles will determine the actual duration of the corrosion process.

Of these factors, relative humidity is the most important. Rapid acceleration of corrosion occurs beyond a critical value of relative humidity [Ju, 1987]. For example, an extreme drop in temperature will cause the sealed cavity to attain its dew point, and condensation will form on the surface of the chip. An electrolyte may form with any ionic contaminants inside the package, serving as an electron transfer medium for the corrosion process. Metal corrosion problems may occur for a variety of reasons in electronic equipment: the use of different metals in electronic products encourages galvanic corrosion, the corrosion resistance of a metal is not constant, and most designers are not aware of corrosion dangers [Sparling, 1967]. Epoxies, plastics, glasses, and ceramics used as insulating or protective overcoats on microelectronic components can provide excellent ionic conduction pathways in the presence or absence of moisture. The properties of the microelectronic product must be understood in order to recognize corrosion as a possible source of failure.

Corrosion mechanisms can be divided into two types: dry, such as surface oxidation of aluminum in air, and wet, in which the reaction occurs in the presence of an electrolyte, a moist environment, and an electromotive force, such as the electrochemical potential difference between two metals. Dry corrosion is of only minor importance in semiconductor products, since the oxidation is self-passivating for the metal, forming a thin oxide film which prevents further oxidation. Wet corrosion in the presence of an ionic contaminant and moisture can provide a conductive path for electrical leakage between adjacent conductors or encourage corrosion of product metallization or bond pads. In wet corrosion, the corrosion product is washed off, exposing fresh metal for further corrosion. Detailed descriptions of various wet corrosion processes (galvanic, pitting, and crevice corrosion) are given in later sections.

Because metals are not completely homogeneous, some areas, such as grain boundaries, are more susceptible to corrosion than others. Corrosion that occurs along grain boundaries is referred to as intergranular corrosion and is usually difficult to detect in the early stages [Sparling, 1967]. In aluminum alloys, the grain boundaries are anodic to grain centers. In a damp environment, pitting can occur preferentially along grain boundaries. Intergranular corrosion is more dominant in marine environments [Davis, 1987].

4.1 CORROSION DUE TO MOISTURE INGRESS

The time to corrosion failure of a package is dependent on the time required for moisture ingress and the time it takes for the corrosion process to cause damage. Corrosion typically begins when the temperature inside the package falls below the dew point, allowing condensation inside the package. Pecht [1990] modeled the time to failure due to corrosion, t_f, as a sum of the moisture ingress time, t_i, and the time for corrosion attack, t_c:

$$t_f = t_i + t_c \tag{4.1}$$

$$t_i = -\frac{4L^2}{\pi^2 D} \ln\left[1 - \left(\frac{P_{in}}{P_{out}}\right)\right] \tag{4.2}$$

where L is the length of the conductor edge exposed to the electrolyte, D is the permeation constant, P_{in} is the partial vapor pressure inside the package, and P_{out} is the partial vapor pressure outside the package. The corrosion time depends on the specific corrosion mechanism.

A package is considered hermetic if its leak rate is less than a specified value. Maximum military leak rates range from 1×10^{-8} atm cc/s, for package volumes less than 0.01 cc, to 5×10^{-7} atm cc/s for package volumes less than 0.04 cc. Using DerMarderosian's data [1988], the moisture ingress time (time to reach three monolayers of water) is determined from Figure 4.1. For example, consider the next equations for leak rate equal to 1×10^{-7} atm cc/s, where V is the volume of the package in cubic centimeters as shown in Table 4.1.

Table 4.1 The Leak Rate as a Result of Volume.

If $0.001 < V < 0.01$	$t_i = V^{.125}\ 10^{3.153}$
If $0.01 < V < 100$	$t_i = V^{.609}\ 10^{4.096}$

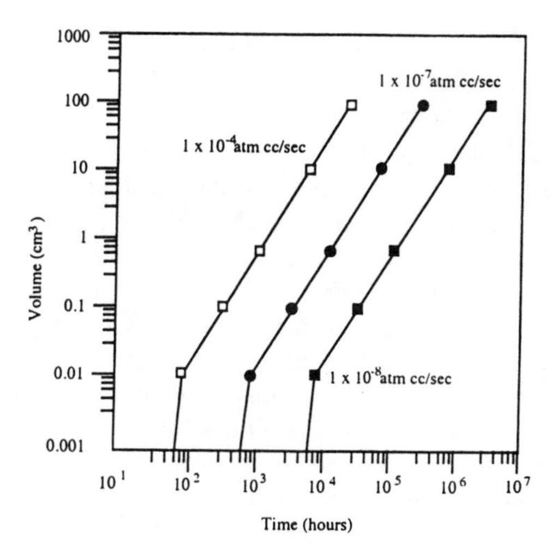

Figure 4.1 Time to reach three monolayers of water as a function of internal volume and leak rate.

This is a worst-case estimate valid at the time of the test, but may prove to be an underestimate if leak paths grow as the package is subjected to further stresses in storage [Pecht, 1990].

Once the critical moisture content is reached inside the package, corrosion can initiate when the non-operating sealed package is exposed to temperatures below the dew point and moisture inside the package condenses and combines with an ionic contaminant (alkali, halides) to provide a conductive path between adjacent metallic conductors. The conductive path serves as a medium for transferring ions during the corrosion process. When the package is in storage without the benefit of operating heat to dissipate the electrolyte, the electrolyte continues to provide a conductive path for the corrosion, which occurs continuously.

Sinclair [1987] of AT&T Bell Laboratories has determined the time to failure for conductors of various cross-sections, given the localized corrosion current through the conductor. His work indicates that a corrosion current of 10^{-13} A has little effect on standard printed circuit board conductor stripes after hundreds of years, but can cause failure of 0.1-µm-wide conductors in tens of seconds.

Peck [1986] has proposed an equation to predict product time-to-failure due to corrosion under accelerated conditions:

$$t_f = A\ (RH)^n \exp\left(\frac{E}{k_B T}\right) \qquad 4.3$$

Peck's research determined that the constants n = -3 and E = 0.9 eV provide a good correlation between his data and times to failure. The experimental constant A is specific to each system and is input by the user. For relative humidity and temperature, the maximum values of a particular environment are used. This equation can be used for normal environmental conditions that reflect accelerated factors, such as the high temperatures and relative humidities of the tropics. On the other hand, the equation cannot be used to predict corrosion failures for sub-zero temperatures and relative humidities below 30%, conditions at which corrosion effects diminish more rapidly than the model predicts.

Typical testing conditions are T = 85°C and RH = 85%; however, to ensure a product has a 20 year life at 30°C and 50% RH, a test at 85°C/85% would have to be run for 5,400 hours, an unreasonable period. Improved test conditions would be 130°C/85% RH for 54 hours, a more realistic duration. Comparison of observed lives with predicted lives has shown a good correlation both above and below the 85°C/85% RH test conditions. Existing data show that life expectations at low stresses can be extrapolated from accelerated testing at high stresses. More tests should be conducted both below and above 85°C/85% RH test conditions: low-stress tests improve statistical results, and high-stress tests improve repeatability.

Higher temperatures have the advantage of reducing testing time; however, industry does not test with temperatures above 140°C to avoid encountering the glass transition temperature, which may decrease with increasing temperature. Furthermore, although 100% RH is useful for reducing test duration, 100% RH is difficult to measure and maintain; 100% RH is thus usually avoided to improve results and optimize extrapolation to low stress levels. Nevertheless, higher stress levels than 85°C/85% RH must be used to reduce unreasonable test times for examining long-term effects [Peck, 1986]. Table 4.2 provides other test conditions, with their acceleration factors in test durations over 85°C/85% RH.

4.2 LOSS OF HERMETICITY

There are three different package types: hermetic packages of metal or ceramic with metal or glass seal rings, quasi-hermetic packages with polymer seals, and plastic packages. Hermeticity is often required for long-term dormancy of sensitive, high-reliability electronics, but it can be lost due to leaks or package deterioration as a result of stress, corrosion, cracking leads, or lid seals. Areas prone to leaks are usually those places where dissimilar materials are joined, such as metal-ceramic interfaces and glass seals embedded in a ceramic or metal package carrier. In hermetic package technologies, the exposed electrical connectors and their underlying protective metallization are in the most jeopardy. If the protective layers are too thin or contain gross defects, corrosion can occur. Hermetic packages can acquire contaminants from inadequate processing or sealing, and from conformal coatings that contain epoxy adhesives. If all water is not removed prior to sealing, even secure, hermetic, ceramic packages can contain enough water to trigger the corrosion process. Ceramic substrates must be subjected to a temperature of 125°C for 48 hours in a vacuum in order to drive out all moisture. If water is present in the package, elevated temperatures drive the moisture out of the epoxy and ceramic substrate to condense on internal package surfaces upon cooling. Moisture trapped beneath the conformal coating on a printed circuit board can react with residual soldering flux to form an electrolyte or a dissolution cell between metal runs or plated through-holes on the board, or the moisture can form phosphoric acid [Devanay, 1989].

Metals and ceramics are very much more resistant than plastics to moisture ingress. Required by military specifications for hermetic packages, ceramics can be very reliably sealed to metal conductors through the use of glass and solders. Yet, ceramic surfaces can still readily attract and retain atmospheric moisture. With the advent of new Novolac injection modeling thermoplastics and the associated reduction in package parts,

Table 4.2 Test Conditions With Acceleration Factors in Test Durations Over 85°C/85% RH.

Condition	Acceleration
121°C/100% RH	16
135/94	30
140/94	40
140/100	50
150/100	77

package fabrication processes, and package interfaces, plastic is a potentially superior packaging alternative. Furthermore, the incorporation of water absorbers, which are used to alter the rate at which water vapor condenses on the package circuitry, is leading to the replacement of traditional packaging methods.

Polymers are mechanically resilient and can be molded to a specified shape. Because they are substantially cheaper and easier to work with than ceramics, epoxy formulations are widely used to encapsulate high production-volume integrated-circuit chips, despite their low thermal stability, modest chemical resistance, and tendency to absorb water. In contemporary integrated plastic packages, the product circuitry is effectively protected if the passivation is free of such flaws as pinholes. The only possible corrosion sites are those parts of the aluminum bond pads left uncovered and unprotected by the gold wire thermosonically bonded to the pad. Since moisture will eventually reach the product by bulk absorption and diffusion, plastic package protection focuses on controlling the kinetics of the ingress of corrosives [Kinsman, 1987]. Product geometry, fabrication techniques, and the composition of the plastic influence contaminant ingress.

The numerous polymer/metal interfaces in packages provide ample opportunities for moisture ingress [Howard, 1987]. Factors that create a moisture pathway between the base metal and the environment at the lead-glass interface include formation of glass meniscus cracks and chipouts after plating, lack of bonding between glass and plating, plating that pulls away from the glass when the leads are bent, and crevices in the lead at the glass seal due to chemical removal of excess oxide after glass sealing. In addition to inhibiting complete plating coverage in the glass seal area, the undercuts can also act as sites for trapping residual plating chemicals and other contaminants that encourage the corrosion process. Corrosion is then accelerated by stress corrosion, over-oxidation of base metal, trapped contaminants, and the electrochemical potential difference between the plating and the lead.

Prevention of lead corrosion at glass-to-metal seals has been addressed in numerous ways. First, organic coatings were applied to the seal area; however, the coatings trapped contaminants when they migrated through poor glass seals and caused internal aluminum bond-wire failures. Secondly, to reduce meniscus cracking and chipouts, leads have been used that have a wider diameter in the glass seal area than above the glass seal, because they move the stress concentration away from the meniscus. Finally, an alternative to meniscus control is to coat the entire lead with nickel prior to glass sealing so that the lead is not dependent on stress allocation for protection of the meniscus. However, tests show that nickel is not completely successful in preventing corrosion, possibly because of iron diffusion throughout the nickel plating. Lead finishes other than nickel have been proposed to reduce corrosion at the glass seal. In tests of various lead finishes, tin was most successful in preventing corrosion, followed by tin/nickel. The gold/nickel finish, on the other hand, did not protect well. Localized base metal attack occurred only in samples that had a gold lead finish. The greater corrosion resistance of tin plating originates from the reduced electrochemical potential difference between the base metal and the lead finish. The most effective form of corrosion protection at the lead-glass interface is tin plating [Berry, 1987].

Since chloride ions and other contaminants can migrate up the plastic leadframe interface and along the gold bonding wire to the bond pad, as diagrammed in Figure 4.2, a key design criterion is to match the thermal expansion coefficients and relative adherence at these interfaces. Focusing on these goals, contemporary molded-plastic packages sustain very high reliability through innovative design and choice of materials [Kinsman, 1987].

Accelerated stress conditions, autoclave tests, and controlled corrosive chemical ambients have been used to assess the susceptibility of plastic-encapsulated products to corrosion. Failure rates under accelerated conditions are determined by the metal deposition process, passivation layer composition and integrity, surface contamination level, leadframe geometry, plastic encapsulant, molding conditions, thermal history, and plastic curing procedures [Schnabble, 1979].

4.3 CORROSION DUE TO CONTAMINANTS

Studies indicate that most contaminants found on electronic equipment in normal environments are particles, rather than corrosive gases like hydrochloric acid and reduced sulfur gases. Sinclair [1987] groups environmental particulates into two categories: particles larger than 15 μm

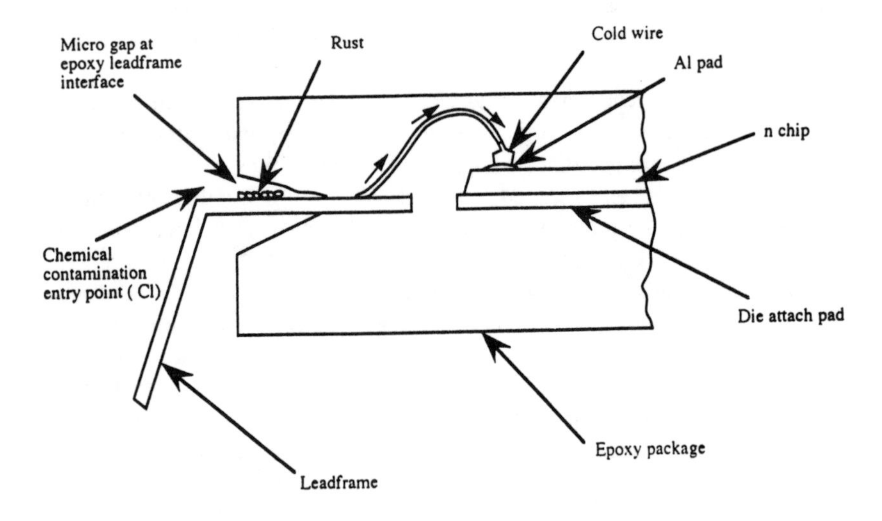

Figure 4.2 Schematic illustration of the interfacial contamination entry point and migration path in molded plastic packages

in aerodynamic diameter (calcium) that can be filtered, and particles smaller than 0.1 μm (sulfate, nitrate, nitrite, ammonium) that enter electronic equipment to settle on vertical as well as horizontal surfaces. Unfortunately, fine particles are rich in ammonium acid sulfate, a very corrosive agent, while calcium-rich coarse particles are less corrosive. Corrosive ammonium acid sulfate particles, commonly present in the air, cannot be filtered [Sinclair, 1987].

The ions most commonly found on the die surface and most likely to form electrolytic solutions, triggering the corrosion process, include phosphorus; halides (especially Cl^- and Br^-); alkali (Na^+) emitted from diffusion ovens, glass containers, and the hands of humans; and sulfur emitted from rubber bands and cardboard separators used to bundle components in storage, which can cause sulfide corrosion of exposed copper-silver eutectic braze alloys [Cieslak, 1987]. Alkali contamination of the surface of the passivation layer is a factor in cathodic corrosion of aluminum-metallized products. The alkali permits the pH at the cathode to locally increase to values in excess of 8. The pH of the electrolyte is an important factor in the corrosion of both gold and aluminum [Schnabble, 1979].

Phosphorus (2 to 4%) is incorporated in the surface passivation glass to prevent cracking during thermal exposure and to restrict the effects of sodium ions. Phosphorus oxides are emitted from phosphosilicate glass, which incorporates phosphorus to lower the melting point for improved reflow; in normal processing, phosphorus oxide is only present in controlled, limited quantities, but process irregularities can lead to areas of high surface phosphorus concentrations [Wagner, 1987]. The phosphorus concentration in phosphosilicate layers used for passivation of aluminum-metallized integrated circuits is a crucial factor in tensile stress and increased susceptibility of exposed aluminum metal to cathodic corrosion. If moisture is present to make phosphoric acid, the glass layer delaminates and the underlying metallization corrodes.

Of all contaminants, chloride ions are the most significant source of corrosion, readily combining with the most common metals (copper, lead, tin) used in electronics. Through an autocatalytic corrosion process, metal chlorides continuously regenerate free chloride ions to sustain the corrosion process [Chen, 1987]. Very little chloride is required to cause considerable damage to the chip. Anodic corrosion of aluminum metallization is frequently observed in chloride contamination. Unfortunately, chloride ions are common among the chemicals used in processing electronic components; chloride ions are a by-product of the epoxide resin chemistry of encapsulation compounds, and can be carried as a trace impurity. Yet, because modern high-performance molding epoxies are specially formulated to reduce hydrolyzable chloride to extremely low levels and getter the remainder into stable compounds, external chlorine sources have become the primary concern. Solder fluxes used on the leads and in the circuit-board soldering process are the primary sources of chloride ions that enter through bond-pad openings and pinholes in the passivation. Processing liquids penetrate defect sites by capillary action and are difficult to remove in subsequent rinsing operations, resulting in significantly higher contaminant concentrations at defect sites.

Atmospheric pollution (sulfur dioxide, nitrite, nitrate, hydrogen sulfide, chloride, salt, ozone) increases the conductivity of the moisture film on the metal surface. In addition, certain pollutants, such as chlorides, reduce the corrosion resistance of passivating films. In industrial areas where coal is burned, the coal releases sulfur dioxide which, in contact with moist air, is converted into sulfuric acid. This acid rain contains as much as 0.01% sulfuric acid. However, these pollutants affect corrosion only when the relative humidity is high. At relative humidities of less than 60% there is no visible change, whether pollution is present or not. In a critical range of 60-80% RH, some corrosion occurs and the protective film breaks down. Above this critical range, there is a large increase in the rate of corrosion if traces of pollution are present; but in pure air, the increase in corrosion is almost negligible, even at 90% RH. The more pollutants are present, the higher is the corrosion [Dummer, 1962]. The critical value of relative humidity beyond which corrosion is accelerated decreases as pollutant concentration increases [Ju, 1987].

4.4 CORROSION DUE TO DEFECTS IN PASSIVATION

Except for bonding pads, thin-film metal structures are almost always coated with an inorganic glassy and/or organic polymeric protective insulating layer that provides both mechanical and chemical protection. These passivating films, if defect-free, can be very effective in preventing corrosion of underlying thin-film conductors. However, localized structural defects in the protective coatings, or the presence of unprotected metal areas such as bonding pads, renders the metal vulnerable to corrosive attack by chemicals from the ambient, such as a plastic encapsulant or in a defective hermetic product atmosphere.

Structural defects in inorganic glassy passivation layers include cracks, pinholes, inadequate edge coverage of metal lines, and other localized defects. Cracks originate from excessive tensile stress in the deposited film, as well as from the tensile stresses of thermal mismatches that develop between aluminum and gold and between phosphosilicate glass and plasma-deposited nitride during the elevated temperatures of product processing. Cracks can also be caused by the recrystallization of aluminum at elevated temperatures (particularly over large metal areas), by contact printing, and by impact during chip handling. Pinholes in dielectric films can be formed during the process of etching the bonding pads. These may result from pinholes in the resist film, inadequate coverage of the resist over edges of topological steps, dust or defects in the photomask, wafer handling, or contact printing. In regions where the gold metallization lines pass over rough surfaces in the underlying dielectric, tensile stresses in the deposited passivation film can cause localized lifting of the passivation film.

In many cases, products fail because of the corrosion of aluminum in regions near bonding pads. In such products, localized defects in the passivation glass provide sites at which corrosion begins. The exposed metal at the defect site serves as one electrode, and an adjacent bonding pad serves as an electrode of opposite polarity. Since the defects in the glass are small, while adjacent exposed metal in the bond-pad area is large, current density at the exposed defect point is high. At such high current densities, localized pH increases at the cathode can result in appreciable dissolution of aluminum [Schnabble, 1979].

The importance of passivation in the metallization corrosion dependence on temperature has been evaluated by Commizolli [1980], who found that the dependence of corrosion current on temperature appeared linear between 60 and 100°C. The corrosion current at 90% relative humidity, using passivated chips, was found to decrease with a decrease in temperature between 60 and 100°C. Generally, corrosion doubles for every 10°C increase in temperature [Sparling, 1967]. Environments that experience high humidity, such as the tropics, inherently have moderately high temperatures — two conditions required for maximum corrosion; therefore, corrosion is high in such areas.

4.5 GALVANIC CORROSION

Galvanic corrosion arises from an extensive use of precious metals (gold, palladium, silver) coupled with base metals (aluminum) in chips and packages [Howard, 1987]. Galvanic corrosion, a form of wet corrosion, occurs when two or more different metals, such as aluminum and gold, are in contact — for example, at a bonding pad — in the presence of an electrolyte. Each metal is associated with a unique electrochemical potential. When two metals are in contact, the metal with the higher electrochemical potential becomes the cathode and the other becomes the anode. The anode is the active metal, where most corrosion occurs; the cathode is the noble metal, which is more corrosion-resistant. The electrolyte can be moisture or a combination of moisture and contaminants, such as ocean salt spray. The electrical contact between dissimilar metals leads to the formation of a galvanic cell. Current flow, driven from the anode to the cathode by the potential difference between the two metals, liberates hydrogen and forms alkali at the cathode.

The rate of galvanic corrosion is governed by the rate of ionization at the anode (the rate at which anode material passes into solution), and this in turn depends on the difference in electrochemical potential between the two metals in contact. The larger this potential difference, the higher the rate of galvanic corrosion. For example, the electrochemical potential with respect to a normal hydrogen electrode is $E_H = -0.76$ for a gold and zinc galvanic cell, and $E_H = 0.34\text{-}0.52$ for a gold and copper cell. The rate of galvanic corrosion will be higher for the gold-zinc pair, all other factors remaining constant. Another example is the use of gold as a lead finish — the large electrochemical potential difference between gold and Kovar increases the rate of galvanic corrosion [Berry, 1987].

The conductivity of the corrosion medium will affect both the rate and the distribution of galvanic attack. In solutions of high conductivity, the corrosion of the more active anodic alloy will be dispersed over a relatively large area. In solutions with low conductivity, most of the galvanic attack will occur near the point of electrical contact between the metals. Yet, even with the most incompatible metals, direct galvanic effects will not extend more than 5 mm from the contact area. When the surface area of the cathode is large in comparison with the area of the anode, the anodic current density is very large and galvanic corrosion is accelerated; when the area ratios are reversed, galvanic corrosion is not as pronounced. For example, when a very small anode of exposed base metal at the lead is in contact with a very large cathode of gold plating, galvanic corrosion occurs rapidly. Larger areas of exposed base metal at the leads corrode at a moderate rate. Furthermore, the closer the physical proximity of the two metals, the higher the rate of galvanic corrosion [Davis, 1987]. Galvanic corrosion eventually ends in failure as an electrical open.

The rate of galvanic corrosion in the presence of a liquid electrolyte increases as temperature increases, due to a more rapid rate of electron transfer. Typically, corrosion products in microelectronic products, such as aluminum hydroxide, are derived from reaction processes that are monotonically increasing functions of temperature. However, temperature is not the most significant factor. Although the corrosion rate depends in part on the steady-state temperature, it also depends on the magnitude and polarity of the galvanic potential, which are functions of electrolyte concentration, pH, local flow conditions, and aeration.

Galvanic corrosion can be minimized by metal-spraying the assembly or electroplating both metals before assembly. The cathodic metal can be electroplated with an anodic finish to reduce the potential difference between the two metals [Dummer, 1962].

The galvanic series — metals arranged in order, from most anodic to most cathodic in a specific electrolyte — identifies which metal or alloy in a galvanic couple is more active. The separation between the two metals or alloys in the galvanic series gives an indication of the probable magnitude of corrosion [Davis, 1987]. Metals do not have a constant electrochemical potential because the potential is affected by environmental factors; therefore, each galvanic series pertains only to a particular environment. A galvanic series can be formed by evaluating the corrosion of a metal when coupled to other metals, and arranging the metals according to increasing corrosion [Kucera, 1982]. Materials can also be selected in accordance with MIL-STD-1250, Table III, Galvanic Couples [Noon, 1987].

4.6 CREVICE CORROSION

Crevice corrosion is a localized attack in a crevice between two metals or a metallic and a nonmetallic surface. One side of the crevice must be exposed to the corroding environment, and the corroding ions must enter the crevice. Crevice corrosion can be very destructive, since the damage is localized and unexpected and since it occurs on metals ordinarily considered corrosion resistant. The mechanism of crevice corrosion is the dissolution of metal M and the reduction of oxygen to hydroxide ions.

Oxygen is the governing factor in these reactions. Initially, the reactions occur uniformly over the entire surface as charge conservation is maintained in both the metal and the solution. After a short interval, however, the oxygen in some crevices is depleted because of restricted convection, and oxygen reduction stops in these areas. Although no further oxygen reduction occurs, the dissolution of metal M continues, producing an excess positive charge in the solution. To balance the charge, chloride ions migrate into the crevice, resulting in an increased concentration of chloride ions there. As a result of hydrolysis, chloride acid is formed, accelerating the corrosion rate as more fresh metal is exposed to attack. Crevice corrosion causes failure rapidly because of the autocatalytic nature of the process.

4.7 PITTING CORROSION

Similar to crevice corrosion, pitting corrosion is a localized attack characterized by surface cavities. It is one of the most destructive forms of corrosion, since only a small percent weight loss is necessary to cause failure. The conditions produced inside the pit accelerate the corrosion process. As the positive ions at the anode go into solution, they become hydrolyzed and produce hydrogen ions. The increased acidity in the pit destroys the adhering corrosion products, exposing more fresh metal to attack. Since the oxygen availability in the pit is low, the cathodic reduction reaction can occur only at the mouth of the pit, limiting lateral growth of the pit [Pecht, 1990]. This autocatalytic process produces porosity, reducing the strength of the material. Chloride ions enhance pitting corrosion [Ju, 1987].

4.8 SURFACE OXIDATION

Surface oxidation, another common type of corrosion in metallic materials, is governed by the free energy of oxide formation. For example, there is a large driving force for the oxidation of aluminum and magnesium, but a smaller oxidation force for copper, chromium, and nickel.

The oxide formed occupies a smaller volume than the metal from which it is formed; tensile stresses can develop in the oxide film, causing it to crack and become porous. Oxidation of subsurface layers can then continue rapidly. In aluminum, silicon, copper, and nickel, the volumes of metal and oxide are relatively equal and an adherent, non-porous, protective oxide film forms, arresting further oxidation. However, the protective oxidation layer can also impair performance in interconnections by reducing electrical conductivity at the surface. If the oxide volume is greater than that of the metal, as with iron, initially a protective layer is formed, but as the layer becomes thicker, high compressive stresses may develop, causing flaking from the surface and exposing subsurface layers of metal for further oxidation, until eventual erosion of the metal.

4.9 STRESS CORROSION

Stress corrosion cracking is an interaction between the two mechanisms of fracture and corrosion, which occur because of simultaneous mechanical stress and corrosion. It results from a stress concentration, applied or residual, at corrosion-generated surface flaws (as quantified by the stress intensity factor, *K*). When a critical value of stress concentration, K_{crit}, is reached, mechanical fracture occurs [Davis, 1987]. Although stress concentrations occur at such flaws, in an inert environment these do not exceed the critical value required to cause mechanical fracture of the material. Thus, stress corrosion essentially reduces the fracture strength of the material so that failure occurs before K_{crit} is reached. The process is synergistic, in that the combined simultaneous interaction of mechanical and chemical forces results in crack propagation, whereas neither factor acting independently or alternately would produce the same result [Davis, 1987].

Stress corrosion differs from other types of corrosion in several respects: tensile stress is required, brittle cracks occur even in ductile metals, and the corrosivity of the environment is not the governing factor. Transgranular or intergranular cracks propagate in a plane normal to the tensile stress — for example, perpendicular to the axis of the leads. In samples exhibiting severe corrosion, base-metal attack causes the leads to break as a result of the 90° bend [Berry, 1987]. Stress corrosion cracking can occur in package leads even without external loads because residual stresses from conventional manufacturing practices are sufficient to initiate the attack. Residual stresses originate during fabrication processes such as rapid-quenching heat treatments. For example, laboratory tests show that failure due to stress corrosion cracking is significantly more frequent at the lead-glass interface than at other areas of the lead because of the presence of large residual stresses originating from the formation of the glass seal, as well as moisture retention at the lead-glass interface. When materials are under a load or stress, even a mild corrosive can cause failure.

Most chemically vapor-deposited silicon dioxide or phosphosilicate passivation glass films are in tensile stress at room temperature. When the passivation layer strongly adheres to the underlying metal, the tensile stress is uniformly distributed. If, however, a crack forms in the dielectric, then the tensile stress in the passivation layer exerts a localized tensile force on the underlying metal at the bottom of the crack, possibly resulting in stress corrosion. Both aluminum and gold films on integrated circuits are in tension at room temperatures. Since aluminum metal is known to undergo stress corrosion cracking, tensile stress in the aluminum film can be a factor in establishing the rate at which an aluminum line corrodes to an electrical open. Tummala [1989] found that electrochemical corrosion interacting with mechanical stress is a potential cause of failure. Failures are typically transgranular and result from the acceleration of the fatigue process by corrosion of the advancing fatigue crack. For example, craze cracking of polymers occurs in halogenated solvent vapors, and adsorbed moisture films due to relative humidity exposure cause static fatigue of the lead zinc-borate sealing glass [Tummala, 1989]. The particular mechanism is chemical attack on glass.

Laboratory tests show that Kovar has a tendency toward stress-corrosion cracking, particularly in the presence of chloride ions. Stress-corrosion failure of a Kovar lead, evidenced by brownish iron rust and intergranular fractures with secondary cracks, results in the separation of the lead at the point of stress. Over-oxidation of Kovar prior to or during glass sealing may contribute to a rapid, localized attack on the lead. Chemical inhomogeneity caused by residual oxide at the grain boundaries increases Kovar's susceptibility to corrosion [Berry, 1987].

Rosengarth [1984] performed numerous stress-corrosion cracking tests on alloys F-15 and 42, which are frequently used for external leads on semiconductor product packages. Good corrosion resistance was exhibited by the alloys in pure water environments at elevated temperature and relative humidity conditions. Stress-corrosion cracking also did not occur in alloys continuously immersed in water containing 3.5% sodium chloride. However, stress-corrosion cracking appeared in alloys that were repeatedly immersed in the salt solution and allowed to dry, possibly because of higher dissolved oxygen concentrations for oxygen reduction and higher chloride concentrations due to evaporation in air. Based on time-to-cracking data, tests in the presence of sodium chloride at an elevated temperature of 70°C with 100% RH caused the most severe stress-corrosion cracking, followed by the alternate immersion test. Chlorine contamination was a necessary condition for stress-corrosion cracking of alloys 42 and F-15, and higher temperatures increased cracking.

Various methods guard against stress corrosion. First, stress corrosion can be avoided by proper material selection. Second, stress-relief annealing can lower residual stress below the threshold. Third, unlike electrodeposited copper and nickel coatings, which are very porous and allow localized corrosion, solder dipping prevents stress-corrosion cracking of the leads by providing a continuous coating [Rosengarth, 1984].

4.10 CORROSION DUE TO MICROORGANISMS

The ideal conditions for maximum growth of most microorganisms are temperatures between 20 and 40°C, with a relative humidity of 85 to 100%. Molds, bacteria, and similar organisms must have water, hydrogen, and oxygen for nourishment. The increasing complexity of synthetic materials makes it difficult or impossible for the designer to decide from the name of a circuit-board laminate or encapsulating resin whether it will support fungus. Many otherwise resistant synthetics are rendered susceptible to microbial deterioration by the plasticizers or hardeners used. The environment, the type of microorganism, and the size, shape, smoothness, and cleanliness of the surface influence the degree of attack.

Acids produced by microbial metabolic processes are corrosive to metals and glass, as well as to organic and synthetic materials. As microorganisms grow, they form an expanding mass that can rupture, loosen, crack, or blister any protective coating. A semi-permeable capsule of microorganisms under a blister produces a local corrosion cell that accelerates metal corrosion. Molds can also upset the electrical balance at the surface of metals, and thus remove the passive film that resists corrosion. It was formerly thought fungal attack could be prevented with a moisture-resistant coating; however, some such coatings are attacked by mold, bacteria, or other microbes, especially if the surface is contaminated by dust or other airborne particles [Sparling, 1967].

4.11 CORROSION PROTECTION

The earth's environment contains numerous deteriorators: oxygen, carbon dioxide, nitrogen, snow, ice, sand, dust, water salt spray, organic matter, and chemicals. Preventing corrosion involves controlling processing and storage environments, using corrosion-resistant materials, improving manufacturing processes and design, and reducing mechanical loads. Equipment specifications often require demonstration that the equipment will withstand specified levels of temperature, humidity, altitude, salt spray, fungus, sunshine, rain, sand, and dust. A material or structure can chemically change in a number of ways. Among them are interactions with other materials (i.e., metal migration, diffusion) and modifications in the material itself (recrystallization, stress relaxation, phase changes, or changes induced by irradiation).

4.11.1 Materials

Chloride and other ionic contaminants are a critical source of corrosion in microelectronics. The presence of these ions is typically reduced in the materials and the processing.

Corrosion-resistant materials should be used as much as possible. Although aluminum is excellent in this respect due to the thin oxide coating which forms when it is exposed to the atmosphere, it tends to pit in moist atmospheres. Furthermore, aluminum may corrode seriously in a salt-laden marine atmosphere. Either anodization or priming with zinc chromate can provide additional protection.

Magnesium is sometimes used to minimize the weight of equipments, but must be protected by surface additives against corrosion and electrolytic reaction. A coating of zinc chromate serves both purposes. Because magnesium is the most reactive metal normally used for structural purposes, the grounding of the magnesium structure requires careful selection of another metal for making the connection. Zinc- or cadmium-plated steel are the more commonly chosen connective materials.

Copper, when pure, is quite resistant to corrosion. However, under certain conditions copper-made parts will become chemically unstable. Transformers have occasionally failed due to impurities in the insulation, and moisture. These conditions favor electrolytic reaction, which causes the copper to dissolve and eventually results in an open circuit.

Iron and steel are used in many parts and structures because of their good magnetic and structural properties; however, only certain stainless steels are reasonably resistant to corrosion. Various types of surface plating are often added, but since thin coatings are quite porous, undercoatings of another metal such as copper are often used as a moisture barrier. Cadmium and zinc plating in some marine environments corrode readily, and often grow "whiskers" which can cause short circuits in electronic equipment.

Materials widely separated in the electromotive series are subject to galvanic action, which occurs when two dissimilar metals are in contact in a liquid medium. The most active metal dissolves, hydrogen is released, and an electric current flows from one metal to the other. Coatings of zinc are often applied to iron so that the zinc, which is more active, will dissolve and protect the iron (a process known as *galvanization*). Galvanic action is known to occur within the same piece of metal if one portion of the metal is under stress and has a higher free-energy level than the other. The part under stress will dissolve if a suitable liquid medium is present. Stress-corrosion cracking occurs in certain magnesium alloys, stainless steels, brass, and aluminum alloys. It has also been found that a given metal will corrode much more rapidly under conditions of repeated stress than when no stress is applied.

Proper design of an equipment therefore requires trade-offs in:

- Selecting corrosion-resistant materials;
- Specifying protective coatings if required;
- Avoiding use of dissimilar metallic contacts;
- Controlling metallurgical factors to prevent undue internal stress levels;
- Preventing water entrapment;
- Using high-temperature resistance coatings when necessary;
- Controlling the environment through dehydration, rust inhibition, and electrolytic and galvanic protective techniques.

4.11.2 Moisture Control

Moisture is one of the most important chemical factors. Moisture is not simply pure water, but usually is a solution of many impurities which can cause chemical difficulties. In addition to its chemical effects, such as the corrosion of many metals, condensed moisture also acts as a physical agent. For example, mated parts can be locked together when moisture condenses on them and then freezes. Similarly, many materials that are normally pliable at low temperatures become hard and brittle if moisture has been absorbed and subsequently freezes. Condensed moisture acts as a medium for the interaction between many, otherwise relatively inert, materials. Most gases readily dissolve in moisture. The chlorine released by PVC plastic, for example, forms hydrochloric acid when combined with moisture. The volume increase from water freezing (i.e., converting from a fluid to solid state) can also physically separate components, materials, or connections.

Although the presence of moisture may cause deterioration, the absence of moisture also may cause reliability problems. The useful properties of many nonmetallic materials such as leather and paper, which become brittle and crack when they are very dry, depend upon an optimum level of moisture. Similarly, fabrics wear out at an increasing rate as moisture levels are lowered and fibers become dry and brittle. Environmental dust can cause increased wear, friction, and clogged filters due to lack of moisture.

Moisture, in conjunction with other environmental factors, creates difficulties that may not be characteristic of the factors acting alone. For example, abrasive dust and grit, which would otherwise escape, are trapped by moisture. The permeability (to water vapor) of some plastics (PVC, polystyrene, polyethylene, etc.) is related directly to their temperature. The growth of fungus is enhanced by moisture, as is the galvanic corrosion between dissimilar metals. Some design techniques that can be used separately or combined to counteract the effects of moisture are

- Eliminating moisture traps by providing drainage or air circulation;
- Using desiccant products to remove moisture when air circulation or drainage is not possible;
- Applying protective coatings;
- Providing rounded edges to allow uniform coating of protective material;
- Using materials resistant to moisture effects, fungus, corrosion, etc.;
- Hermetically sealing components, gaskets, and other sealing products;
- Impregnating or encapsulating materials with moisture-resistant waxes, plastics, or varnishes; and

- Separating dissimilar metals, or materials that might combine or react in presence of moisture, or components that might damage protective coatings.

The designer also must consider possible adverse effects caused by specific methods of protection. Hermetic sealing, gaskets, protective coatings, etc., may, for example, aggravate moisture difficulties by sealing moisture inside or contributing to condensation. The gasket materials must be evaluated carefully for outgassing of corrosive volatiles or for incompatibility with adjoining surfaces or protective coatings.

In dormant environments, controlling the relative humidity is often more useful for corrosion resistance than controlling the temperature. Industry-sponsored investigation during the past two years has shown that high corrosion rates occur when there is high humidity cycling, although the median relative humidity is low and the temperature is controlled. Therefore, large magnitudes of relative humidity cycling, rather than high steady-state temperatures, are critical to the corrosion process. Relative humidity should be controlled so that it changes less than 5% in an hour [Noon, 1987].

To predict corrosion resistance, the ten-day moisture-resistance cycling test evaluates the resistance of component parts to the deteriorating effects of high humidity and temperatures typical of the tropics. Unlike a steady-state humidity test, this accelerated test involves temperature cycling to provide the alternate periods of condensation and drying essential to corrosion [Bratschun, 1987].

Tests have also involved the application of contaminants through various methods. Evaporation techniques enable the application of a known concentration of contaminants as one of the constituents of a moist atmosphere. However, most frequently, the product is immersed in a dilute solution, such as aqueous sodium chloride or sodium bicarbonate. The immersion test additionally simulates effects that occur in product processing, when liquids containing contaminants are used for etching, resist stripping, or other operations — the stage at which corrosion is initiated before contact with contaminants in the atmosphere [Schnabble, 1979].

Chapter 5

RADIATION FAILURE MECHANISMS

Electronic products stored near nuclear reactors, isotropic nuclear sources, accelerators, or nuclear detonations must be designed to tolerate the two basic effects of nuclear irradiation — mechanical and electrical failures. The mechanical failure mechanism causes time-dependent wearout failures through an embrittlement phenomenon that increases the hardness and decreases the ductility of metals. However, the electrical failure mechanism — random overstress when a single radiation particle interacts with the LSI/VLSI circuitry — is a more critical problem.

The radiation environment in space near the earth is composed, primarily, of the Van Allen, auroral, solar flare, and cosmic phenomena. Others of lesser importance are solar wind, thermal energy atoms in space, neutrons, naturally occurring radon gas, albedo protons, plasma bremsstrahlung, and man-made nuclear sources. In the electromagnetic spectrum are gamma rays, X-rays, ultraviolet and Lyman-alpha radiation.

In general, metals are quite resistant to radiation damage in the space environment. Semiconductor devices may be affected by gamma rays which increase leakage currents. The lattice structure of semiconductors can be damaged by high-energy electrons, protons, and fast neutrons, which cause permanent effects through atomic displacement and damage to the lattice structure. Organic materials are particularly susceptible to physical changes in cross-linking and scission of molecular bonds. Radiation induced formation of gas, decreased elasticity, and changes in hardness and elongation are some of the predominant changes in plastics which have been subjected to radiation of the type encountered in the space environment. Electronic circuits are affected by a lowering of input and output impedances.

The electromagnetic spectrum can be divided into several categories ranging from gamma rays at the short-wavelength end through X-rays, ultraviolet, visible, infrared, and radio, to the long-wavelength radiation from power lines. Damage near the surface of the earth is caused by the electromagnetic radiation in the wavelength range from approximately 0.15 to 5 m. This range includes the longer ultraviolet rays, visible light, and up to about midpoint in the infrared band. Visible light accounts for roughly one-third of the solar energy falling on the earth, with the rest being in the invisible ultraviolet and infrared ranges. The solar constant (the quantity of radiant solar heat received normally at the outer layer of the atmosphere of the earth) is very roughly about 1 kW/m^2. In some parts of the world, almost this much can fall on a horizontal surface on the ground at noon.

Solar radiation principally causes physical or chemical deterioration of materials due to photo-radiation induced reactions, such as photo-induced degradation of rubber. Radiation can also cause effects, such as the temporary electrical breakdown of semiconductor devices exposed to ionizing radiation. Radiation protection analysis requires knowledge of the irradiated material and its absorption characteristics and sensitivity to specific wavelengths and energy levels, ambient temperature, and proximity of reactive substances such as moisture, ozone, and oxygen. Some specific protection techniques are shielding, exterior surface finishes that will absorb less heat and are less reactive to radiation effects of deterioration, minimizing exposure time to radiation, and removing possibly reactive materials by circulation of air or other fluids or by careful location of product.

Another form of natural electromagnetic radiation is that associated with lightning. It is estimated that lightning strikes the earth about 200 times each second, each stroke releasing large bursts of electromagnetic energy. Most of this energy is concentrated at the low-frequency end of the electromagnetic spectrum, with the maximum power level being concentrated at about 3 kHz.

Man-made electromagnetic energy is another form and is of far greater importance when solar energy is excluded. Artificial electromagnetic radiators include power distribution systems, a multitude of uses in communications, and specialized detection and analytical applications. The development of lasers has introduced another intense source of electromagnetic radiation and, in military application, the electromagnetic pulse (EMP) associated with nuclear weapon detonations is of considerable importance.

The EMP spectrum is similar to that created by lightning, with a maximum energy appearing at about 10 kHz but distributed with smaller amplitudes throughout a broad region of the frequency spectrum. EMP energy is of considerably greater magnitude than that observed in lightning and extends over a much larger area of the earth. Despite the similarities among EMP and lightning and other strong sources of electromagnetic energy, it cannot be assumed that protective measures consistent with these other electromagnetic radiation sources will protect material from the effects of EMP. The rapid rise time of the pulse associated with a nuclear detonation and the strength of the resulting pulse are unique.

While the variety of the effects of electromagnetic radiation on materials are known, models are still being developed, and the effects on man are somewhat poorly understood. One of the most important effects of electromagnetic radiation in the environment is the electromagnetic interference (EMI) it produces on the effective use of the electromagnetic spectrum. Well-known examples are called radio interference and radar clutter. Another important effect in the military is the interaction of electromagnetic radiation with electroexplosive devices used as detonators. Military as well as civilian explosives are provided with detonators that often depend on heating a small bridge wire to initiate the explosion. Absorbed electromagnetic radiation can accidentally activate such fuzes.

Protection against the effect of electromagnetic radiation has become a sophisticated engineering field of electromagnetic compatibility (EMC) design. The most direct approach to protection is, in most cases, to avoid the limited region in which high radiation levels are found. When exposure cannot be avoided, shielding and filtering are important protective measures. In other cases, material design changes or operating procedural changes must be instituted in order to provide protection.

With the increasing use of X-rays, electrostatic generators, cyclotrons, and so forth, more intense sources of radiation have become available. For example, nuclear reactors and atomic discharges provide considerable radiation. Nuclear radiation varies considerably in how it reacts with matter [Jowett, 1973]:

- Changed particles interact through elastic, inelastic, bremsstrahlung and with the nuclei;
- Gamma particles interact with matter through photoelectric, compton, pair-production;
- Neutrons cannot cause excitation or ionization, but they interact with the nuclei of atoms in elastic collisions.

Neutrons and gamma rays cause most of the damage to microelectronics, the reaction of neutrons with matter varies with energy. Products can only be tested by subjecting them to the actual conditions of nuclear radiation and measuring the threshold of damage and tolerance levels. Semiconductor materials, such as germanium and silicon, are also sensitive to radiation, but organic materials are damaged more than inorganic. Most silicon and germanium semiconductors are damaged at neutron levels of 10^{11} to $10^{12}/cm^2$ [Jowett, 1973].

The effects of radiation occur when nuclear radiation in any form interacts with the target. This results in the transfer of energy from the incident nuclear particle to the target atom. This transfer of energy can be manifested as electronic excitation of the atom, production of a free electron with kinetic energy, or creation of new particles. It can also appear as the kinetic energy of the entire atom.

Three general types of radiation effects are displacement effects, transient effects, and radiation-induced chemical effects [Olesan, 1966]. Macroscopic physical properties of the material are impacted by displacement effects, in which atoms are displaced from lattice sites when a fast nuclear particle collides with the nucleus of an atom transferring sufficient energy. The displaced atoms move through the material, losing energy in collisions with other atoms and simultaneously displacing some of them. The resulting lattice defects may anneal or form secondary defects. Transient effects are due to changes in electron states, and produce considerable changes in the electrical and optical properties of the materials. Transient effects are also a problem for electronic circuits, especially military electronics that are either stored or operated in an intense nuclear radiation environment.

In some systems, chemical change is the primary effect of radiation, which interacts with atomic electrons, producing free electrons and positive ions. The positive ions undergo secondary reactions, while the free electrons lose energy by inelastic and elastic scattering. Positive and negative ions recombine to release the kinetic energy of the resultant molecules. Some of the recombination products may be chemically active free radicals that can participate in secondary chemical reactions [Olesan, 1966].

5.1 MECHANICAL DEGRADATION

In metals and ceramics, radiation causes mechanical failures through point defects such as Schottky defects (atoms knocked out of molecular lattice structures and lodged in interstitial sites). During the formation of a point defect, a nuclear particle enters a material and collides with the nucleus of an atom, giving it the energy needed to break away from its position in the lattice. Several factors influence the type of defect formed: the type and energy of incident radiation, the temperature and charge state of defect, and the impurities and lattice imperfections present before radiation. The displaced atom can then move through the lattice, displacing other atoms.

This atom, with less energy than the incident radiation, moves at a lower velocity that usually cannot produce ionization events. The thermal energy of the crystal enables some of the simple defects and defect clusters to migrate through it. The ability of the defect to move is strongly affected by its charge state and temperature. Excess electrons can move under the influence of an electric field; positive charges can move in some solids by capturing an electron from a neighboring atom (hole motion). Surprisingly, an interstitial atom can move about in many solid lattices at room temperature. Similarly, a vacancy can move by capture of an atom from an adjacent lattice site, which then becomes a vacancy. The mobility of the ion is a function of its mass and temperature; a charged defect has greater mobility than a neutral one.

Eventually, mobile defects are annihilated by the recombination of vacancy-interstitial pairs, become immobilized by formation of stable defects with the impurities of lattice defects, or escape to a free surface. Displaced atoms finally come to rest when the energy is fully released, lodging themselves in vacancies or interstitial sites. However, an electron or ion may recombine, possibly disrupting the chemical make-up of a molecular ion. Temperature influences the recombination rate, but immobilization may not be permanent. For example, an electron or hole may be trapped with a low enough binding energy to be released thermally after a while. In other cases, the untrapping can be stimulated by heat, large electric fields, or mechanical stress. Many delayed radiation effects are due to this process.

Although some defects lead to secondary defects, others anneal. During annealing, a crystal with defects is warmed to a temperature at which the defects become mobile and structural rearrangements occur, restoring the crystal to a pre-irradiation state. At low doses of radiation, the lattice structure recovers when displaced atoms fill vacancies. In silicon, interstitial atoms that escape from the vacancy site when mobilized by ionization during irradiation migrate to grain boundaries.

Point defects alter the mechanical, thermal, optical, and electrical properties of the metals. More specifically, they cause embrittlement aging, which can be countered by annealing, increase the electrical resistivity of metals, change the thermal conductivity of materials, and alter the carrier mobility and effective doping of semiconductors [Olesan, 1966]. The contribution made by a defect to a material property change depends strongly on its charge state. For example, a donor or acceptor level changes the majority-carrier concentration only when it captures or contributes a carrier and becomes charged [Van Lint, 1980]. In polymeric materials, radiation aging is caused by breaks in polymer chains or changes in the degree of polymerization due to chain branching. Photodegradation of polymers under prolonged exposure to UV radiation in strong sunlight reduces the strength of the polymer; therefore, stabilizers are sometimes needed to combat such wearout failures.

Fast neutrons cause the most damaging displacement effects. Relatively heavy chargeless particles, neutrons are very penetrating. Generated in immense quantities in nuclear explosions and reactors, neutrons only interact by direct (mostly elastic) collisions with the nuclei of atoms. Neutrons, being uncharged, dispel a large fraction of their energy in atomic displacement — primarily in clusters, a group of closely spaced lattice defects that result from very energetic recoiling atoms after a large amount of energy is transferred to the lattice. The reduction in cluster strength is believed to cause short-term annealing of neutron damage. Annealing of large clusters can produce some simple defects, caused by photons and electrons. The extent to which photons, electrons, neutrons, or any other particles interact with the nuclei is described in terms of quantities known or cross-sections. The cross-section (σ) for interaction between particles (including photons) and a stopping material is defined as

$$\sigma = \frac{P}{\phi} \quad 5.1$$

where p is the probability of the interaction when one target atom or molecule is subjected to the particle (or photons) fluence ϕ. The unit of σ is the barn (b), one barn = 10^{-28} m^2.

The effect of neutrons on semiconductors is more pronounced than that of gamma rays because they damage the crystal lattice, permanently changing the electronic properties of the material. The most important effect of neutron damage on bipolar resistors is the reduction of minority-carrier lifetime, which reduces current gain. While gamma rays can displace atoms from their lattice sites, neutrons displace many more because of their greater effective mass. The resulting defects and band gap traps increase carrier scattering and decrease carrier mobility. Reduced mobility translates to reduced conductivity and frequency response.

Gamma rays (photons) damage indirectly through the generation of Compton electrons, which mainly cause point defects. The lack of mass and charge in electromagnetic waves, such as gamma and X-rays, makes them very penetrating during ionizing radiations (ejection of bound electrons from atoms). Inorganic materials are damaged by the electric charge freed by ionization; during ionization, gamma rays strip electrons from their bound states, leaving behind less mobile holes that become trapped in the gate oxide, creating an unwanted positive charge. This charge lowers the threshold voltage by as much as 30 V. In such instances, MOS devices are more susceptible to surface changes than bipolar devices. Ionizing radiation produces excess electron-hole pairs within all semiconductor materials, resulting in transient leakage photocurrents that appear across reverse-biased semiconductor junctions.

Other high-energy particles causing displacement effects include protons and electrons. Because of its larger mass, a proton in a head-on collision with a nucleus can impart much more energy than an electron. Electrons are the least damaging, primarily causing such point defects as Frenkel defects, in which both a cation and an anion are missing from the lattice structure. Unlike neutron damage, electron damage is impurity-sensitive. Higher-energy electrons (>5 MeV) and protons produce a mixture of simple defects and clusters [Van Lint, 1980].

Displacement effects cause permanent damage to the crystal structures in semiconductors. Radiation environments that contain heavy particles in the flux or a large amount of high-energy photons cause particle interactions with semiconductor nuclei through the production of secondary particles that cause displacements.

In order to determine the time-to-failure due to mechanical degradation, a critical void size specified by military standards can be used as a guideline. The time required for voids in the material to reach the critical size dictates the time to mechanical failure through radiation damage.

5.2 GAMMA AND NEUTRON HEATING

When electronic parts are located near nuclear devices, such as nuclear propulsion systems, the effect of gamma and neutron heating becomes considerable, especially when the nuclear devices are lightly shielded. Gamma and neutron heating is dependent on various parameters, such as total absorbed dose-rate, mass attenuation coefficient, specific heat of the absorber, the gamma neutron spectrum of the reactor, reactor power, and operating time.

Heating is caused by collisions between the nuclei of the material and neutrons, which primarily heat organic materials with a high hydrogen content [Olesan, 1966]. The heat deposition rates for three typical materials are shown in Table 5.1.

These heat deposition rates are based on a relative biological effectiveness (RBE) equal to 1 for gamma radiation and 10 for neutrons, assuming an unshielded reactor with a high number of thermal neutrons in its spectrum. The times needed to achieve the equilibrium temperature vs. distance from the reactor and reactor power are graphed by Olesan [Graphs 6.1, 6.2, 6.3, 6.4].

Table 5.1 Heat Deposition Rate for Polyethylene, Aluminum, and Iron.

Material	Density [g/cm^3]	Heat deposition rate	
		Neutrons	Gamma radiation
Polyethylene	1.0	9.18^{-12}	4.06^{-11}
Aluminum	2.7	2.98^{-13}	1.02^{-10}
Iron	7.8	1.43^{-13}	3.09^{-10}

5.3 INSULATION DEGRADATION

The insulation resistance of certain materials decreases by a factor of from 10 to 100. Polyvinyl chloride and silicon rubbers have poor resistance, while polyethylene can withstand intense radiation for as long as five months before dropping below 5 x 10^9 ohms [Jowett, 1973].

5.4 ELECTRICAL DEGRADATION AND SINGLE-EVENT UPSET

The electrical failure modes caused by radiation dictate, in part, the choice of packaging materials. Radiation effects can be a serious obstacle to further rapid increases in VLSI densities, particularly in memory chips, which usually lead other microelectronics technologies in advanced development. Cosmic rays or high-energy particles (electrons, photons, muons, pions, neutrons, or alpha particles) can cause sudden random electrical failures in an ionization event — known as a single-event upset (SEU) or soft error — by passing through the microcircuit and adding enough charge to surpass the critical charge (100 fC) that represents a bit, thus temporarily changing the logic state. Single-event upsets are a consequence of the evolution of integrated circuits; increased density and speed and decreased power and cost per bit have been accomplished by decreasing both the cell size and the critical charge that represents a bit. As memory sizes increase and memory cell sizes shrink, the number of single-event upsets increases.

Single-event upsets are produced by the direct ionization of a particle as it loses energy passing through the microcircuit, or by the ionization of secondary particles created by medium-energy nuclear reactions. The ionization creates electron-hole pairs. If the charge is generated in the vicinity of a p-n junction with a reverse bias, the intense electric fields collect the charge. The electrons and holes are separated, and the charge of the appropriate sign is collected, while the opposite charge is swept out of the depletion region. The event does not cause any physical defects and the logic state soon returns to normal if irradiation ceases, allowing time for recovery. Since these transient effects are only critical during operation, single-event upsets will not be significant during dormancy. However, radiation can build up during long-term dormancy, causing electrical failures when the product is operated later. Surface layers of transistors and semiconductor devices, when exposed to electromagnetic radiation for a long time, attract gas ions, upsetting their internal electric fields.

5.5 RADIATION DAMAGE DUE TO ALPHA PARTICLES

Of all the radioactive particles, alpha particles are the most critical source of electrical failures due to radiation damage. Naturally occurring alpha-particle radioactivity in packaging materials causes soft-error failures (random, non-recurring single-bit errors) in random access memory semiconductor devices under the following conditions [May, 1979]:

- When alpha-emitting radioactive isotopes are present near the interior of the package (within 50 microns), allowing escape of high-energy alpha particles before they are stored internally;
- When emitted alpha particles have a direct path to the device surface; and
- When the carriers generated by the passage of the particles are collected in sufficient quantity to upset a memory bit.

Uranium and thorium impurities in LSI materials became important when May [1979] discovered that alpha-particle emission from these contaminants was responsible for soft errors in memories through the generation of electron-hole pairs. In silicon, a typical 5-MeV alpha particle can generate 1.4 million electron-hole pairs through a penetration distance of 25 microns. Many naturally occurring materials used to manufacture packages contain at least some measurable impurity level of uranium and thorium, sufficient to cause soft errors. If a thousand 64K RAMs are used in a computer system and the probability of a soft error in a single 64K RAM is one in a million, then the computer can fail as often as once every thousand hours with uranium and thorium concentrations of less than 1 ppm. Elimination of the problem by purifying the LSI materials requires a reduction in the concentration of uranium and/or thorium to three orders of magnitude.

Since plastic, being organic, is relatively free of radioactive species and contributes little to alpha radiation levels, plastic packages can be controlled to some extent. Naturally occurring alumina ores contain 1 to 100 ppm uranium and thorium, which may be reduced by a factor of two during purification. However, the remaining uranium/thorium impurities, present as substitutional atoms for the aluminum, are difficult to remove without expensive chemical processing. Naturally occurring silica, with a uranium/thorium impurity level of a few parts per million, is resistive to purification. Hence, plastic packages that contain 50 to 70% silica or alumina fillers exhibit radioactivity characteristic of uranium/thorium impurity levels of a few ppm. Basically, controlling radiation in plastic packages is accomplished by using fillers with fewer contaminants.

Hermetic packages are more complex. They consist of an alumina package body, lid, and lid seal, all of which may be radioactive and made from a combination of materials (glasses, metals, and ceramics). Trace amounts of thorium or uranium naturally occur in the aluminum oxides and sealing glasses of hermetic packages. Many of the glasses used in glass-sealed hermetic packages also contain up to 20% zircon filler, to aid in controlling thermal expansion. Natural zircon contains 100 to 1000 ppm uranium, a level of radioactivity substantially above that in either alumina or silica. Since the uranium and thorium are bound in the lattice as substitutions for zircon ions, eliminating these impurities without eliminating zircon is complex and expensive.

For hermetic packages, sealing glasses are ten to a hundred times more radioactive than ceramics; however, the geometrical configuration of sealing glasses limits radioactivity to the same level as that of ceramics. Metal-lidded ceramic packages are generally less radioactive than ceramic-lidded packages. Likewise, gold-tin braze alloys are less radioactive than solder glasses. The lid of a hermetic package emits the majority of alpha particles produced.

Radioactive impurities have also been detected in iron and aluminum alloys refined since World War II. Since many commercial alloys and much of the aluminum used for metallization are recycled, the Kovar or alloy 42 used in semiconductor packages can often be contaminated with radioactive impurities. Hence, even metal lids coated with 1 to 3 microns of gold can exhibit alpha radioactivity.

Package manufacturers must try to reduce the already low levels of radiation in package materials, since new products with smaller critical charges need to be more sensitive to alpha radiation. One solution is to interpose a non-radioactive shielding material between the source of radiation and the product. Although shielding is inexpensive and manufacturable, it may itself contain very low levels of radioactive species or introduce other reliability problems. If a shield with residual radioactivity is deposited on the silicon die surface, emitted alpha particles with a direct path to the memory cells will cause severe damage. To be effective, the shield material must be free of any residual radioactivity and also be thick enough to completely bar all alpha particles, not just slow them down or stop only some. To prevent delamination, which can introduce paths for moisture collection, the shield material must exhibit good adhesion to the uppermost layer of the active product. Silicone at least 0.03 mm thick will stop alpha particles in the 8.0 MeV range.

Another solution is to remove radioactive species from the package materials. These are easier to remove from plastic packages, in which only the filler must be considered, than from hermetic packages. The main difficulty of this method is detecting low-level radioactivity, an expensive process. Large-area gas proportional flow counting, sophisticated gamma spectroscopy, liquid scintillation, and nuclear track counting are techniques for measuring low-level radioactivity. Scintillation detectors are easy to use, economical, and efficient; however, they are subject to high background radiation levels. Gas proportional flow detectors are useful because large-area detectors can reduce count time. Nuclear film tracking has the advantage that films can be left unattended for weeks or months. Finally, atmospheric factors, such as airborne radon, may contaminate wafers and open packages during fabrication and measurement, which requires very low background levels in order to resolve low count rates.

5.6 RADIATION SHIELDING

Radiation shielding is necessary to electronic components and products. The two types of shielding, active and passive, reflect the type of damping applied to the motion of the radiation particles. Passive shielding has not been employed in the past, due to the weight of the shielding material, but in the age of shrunken circuit sizes, passive shielding has become more feasible. Active shielding can be either electrostatic or magnetic. Passive shielding is preferred, because encapsulating materials are not dense enough to prevent the passage of neutrons.

The effects of neutron bombardment on electronic components are significant, necessitating shielding. If the initial radiation effect from a pulsed source is intense, gamma radiation may cause transient effects that can damage organic and semiconductor materials due to the ionization of gases. Ionization of air and gases can produce leakage paths. Damage due to air ionization by intense gamma radiation can be prevented by encapsulating the circuits with conformal materials, such as silicone rubber, eliminating air pockets that could support ionization.

Chapter 6

MECHANICAL FAILURE MECHANISMS

Mechanical failure mechanisms which can arise during non-operating are presented in this chapter. The main failure mechanisms include fatigue and fracture of various package components, which occur primarily as a result of mismatches in the coefficients of thermal expansion of the various materials that make up the package. Vibrations and shock can also induce cracks and crack propagation in various package components.

6.1 FAILURES INDUCED BY TEMPERATURE

Table 6.1 summarizes the temperature dependencies of mechanical failure mechanisms. Table 6.2 gives some of the low-temperature concerns and protection methods. Details of the temperature effects on electronics can be found in Pecht [1991].

The die, substrate, leadframe, and case of a microelectronic package typically have different thermal expansion coefficients. For example, dies are usually made of silicon, gallium arsenide, or indium phosphide, while the substrate is typically alumina, beryllia, or copper, with a coefficient of thermal expansion different from the die material. During temperature cycling, tensile stresses develop in the central portion of the die and shear stresses develop at the edges. Ultimate fracture of the brittle die can occur suddenly and without any plastic deformation when surface cracks at the center or edge of the die reach critical size and propagate during thermal cycling.

The stress generated in the die during temperature cycling is dependent on the magnitude of the temperature change. Improper die to header bonds or unsuitable packaging conditions can increase the incidence of die cracks [Tan, 1987]. Wafer back-processing, such as lapping or thinning, can result

Table 6.1 Temperature Dependence of Product Failure Sites.

Failure site	Failure mechanism	Dominant temperature dependence	Nature of steady-state temperature dependence	References for models
Wire	Flexure fatigue	Δ T	Independent of steady-state temperature function under normal operation	[Gaffeny, 1968], [Villela, 1970], [Ravi, 1972], [Phillips, 1974], [Pecht, 1989], [Harman, 1974]
Wirebond	Shear fatigue	Δ T	Independent of steady-state temperature	[Philosky, 1973], [Pecht, 1989], [Newsome, 1976], [Philosky, 1971], [Gerling, 1984], [Khan, 1986], [Pinnel, 1972], [Feinstein, 1979a], [Feinstein, 1979b], [Pitt, 1982]
	Kirkendall voiding	T	Independent of steady-state temperature below 150°C; independent of temperature under 150°C in presence of halogenated compounds	[Newsome, 1976], [Philosky, 1971], [Gerling, 1984], [Khan, 1986], [Pinnel, 1972], [Feinstein, 1979a, 1979b], [Pitt, 1982], [Villela, 1971]
Die	Fracture	Δ T, ∇ T	Primarily dependent on temperature cycle	[Tan, 1987], [Hawkins, 1987]
Die adhesive	Fatigue	Δ T	Independent of steady-state temperature under normal operation	[Chiang, 1984], [Mahalingham, 1984]

Failure site	Failure mechanism	Dominant temperature dependence	Nature of steady-state temperature dependence	References for models
	Cracking	Δ T	Independent of steady-state temperature (T) below the glass-transition temperature of the encapsulant; a Δ T, ∇ T driven mechanism	[Nishimura, 1987], [Fukuzawa, 1985], [Kitano, 1988]
Package	Stress corrosion	dT/dt	Mildly steady-state temperature-dependent under normal operation	[Tummala, 1989]
Die metallization	Corrosion	dT/dt	Only occurs above dew-point temperature. Mildly steady-state temperature-dependent under normal operation	[Pecht, 1990], [Commizolli, 1980] [Inayoski, 1979], [Sim, 1979], [White, 1969], [Schnabble, 1969]
	Constraint cavitation	T	Steady-state temperature dependent above 25°C	[Yost, 1988], [Yost, 1989]
	Electrostatic discharge	T	ESD voltage (i.e., resistance to ESD) reduces with temperature increase (from 25 to 125°C). Not a dominant mechanism in properly protected products	[Kuo, 1983], [Hart, 1980], [Moss, 1982], [Amerasekera, 1986, 1987], [Scherier, 1978]

in silicon flaws that are susceptible to fracture during temperature cycling [Hawkins, 1987]. Die fracture and die adhesion fatigue are dependent primarily on the magnitude of the temperature cycle. Components stored in arctic and desert environments undergo temperature fluctuations of up to 20°C per day; such temperature cycling over a long period of time can eventually result in die fracture failures.

6.2 FAILURE INDUCED BY SHOCK AND VIBRATION

Shock and vibration are common accelerators of failure. The most frequent vibration-induced failures include:

Table 6.2 Low-Temperature Protection Methods.

Effect	Preventive measures
Differential contraction	Careful selection of materials Provision of proper clearance between moving parts Use of spring tensioners and pulleys for control cables Use of compliant material for skins
Lubrication stiffening	Proper choice of lubricants: • Use greases compounded from silicones, diesters or silicon diesters thickened with lithium stearate • Eliminate liquid lubricants wherever possible
Leaks in hydraulic products	Use of low-temperature sealing and packing compounds
Stiffening of hydraulic products	Use of proper low-temperature hydraulic fluids
Ice damage caused by freezing of collected water	Elimination of moisture by • Provision of vents • Eliminating moisture pockets • Suitable heating • Sealing • Desiccation of air
Degradation of material properties and component reliability	Careful selection of materials and components with satisfactory low-temperature capabilities

- Flexing of leads and interconnects;
- The dislodging or damaging of parts and structures, or foreign particles in filters, bearings, pumps, and electronics.

The degree of failure generally depends on the natural frequencies, deflections, and mechanical stresses within components and materials produced by the shock and vibration environment. If the mechanical stresses so produced are below the acceptable long-term safe working stress of the materials involved, failures should not occur. If, on the other hand, the stresses exceed the safe levels, cumulative damage and eventual failure will occur, and corrective measures such as stiffening, reduction of inertia and bending moment effects, and incorporation of further support members are indicated. If such approaches do not reduce the stresses below the acceptable safe levels, further reduction is usually possible by the use of shock absorbing mounts.

Equipment is sometimes specially mounted to counter the destructive effects of shock and vibration. Shock mounts often serve this purpose, but effective means are complex for attenuating shock and vibration simultaneously. Isolation of an equipment against the effects of vibration requires that the natural frequency of the equipment be substantially lower than the undesired frequency of vibration.

Three basic kinds of isolators are available:

- Elastomers made of natural or synthetic rubber, used in a shear mode or in a diaphragm to damp the induced shock or vibration;
- Metallic isolators, which include springs, metal meshes, or wire rope. Springs lack good damping qualities, but meshes and rope provide smooth friction damping;
- Viscous dampers (similar to the type used on automobiles) which are velocity-sensitive, although they tend to become ineffective under high-frequency vibration; resilient mounts must be used with caution. If improperly placed they can amplify the intensity of shock and vibration; the ideal goal is to design equipment to be resistant to shock and vibrations, rather than to provide complete isolation.

Another vibration protective technique is potting or encapsulation of small parts and assemblies. The choice of the material to be used is often critical. The potting material should be compliant enough to dampen vibrations. Some materials polymerize exothermically, and the self-generated heat may cause cracking of the casting or damage to such heat-sensitive parts. Heat sinks or powdered fillers are often used to increase heat conduction. Another problem is shrinkage of the potting material. In some cases, molds are made oversized to compensate for shrinkage.

One factor which is not often considered is that the vibration of two adjacent components, or separately insulated subsystems, can cause a collision between them if maximum excursions and sympathetically induced vibrations are not evaluated by the designer. Another failure mode, fatigue (the tendency for a metal to break under cyclic stressing loads considerably below its tensile strength), includes low cycle fatigue, acoustic fatigue, and fatigue under combined stresses. The interaction between multiaxial fatigue and other environmental factors such as temperature extremes, temperature fluctuations, and corrosion requires careful study. Stress-strength analysis of components and parameter variation analysis are particularly suited to these effects.

In addition to using proper materials and configuration, the shock and vibration experienced by the equipment should be controlled. In some cases, however, even though an item is properly insulated and isolated against shock and vibration damage, repetitive forces may loosen the fastening products. If the fastening products loosen enough to permit additional movement, the product will be subjected to increased forces and may fail. Many specialized self-locking fasteners are commercially available, and fastener manufacturers usually will provide valuable assistance in selecting the best fastening methods.

An isolation product can be used at the source of the shock or vibration, in addition to isolating the protected component. The best results are obtained by using both methods. Damping products are used to reduce peak oscillations, and special stabilizers employed when unstable configurations are involved. Typical examples of dampeners are viscous hysteresis, friction, and air damping. Vibration isolators commonly are identified by their construction and material used for the resilient elements (rubber, coil spring, woven metal mesh, etc.). Shock isolators differ from vibration isolators in that shock requires stiffer spring and a higher natural frequency for the resilient element. Some of the types of isolation mounting products are underneath, over-and-under, and inclined isolators.

The basic considerations when designing for shock and vibration include the location of the component relative to the supporting structure (i.e., at the edge, corner, or center of the supporting structure), the orientation of the part with respect to the anticipated direction of the shock or vibration forces, and the method used to mount the part.

6.3 FAILURE INDUCED BY SAND AND DUST

Sand and dust can degrade equipment by abrasion, leading to increased wear, friction causing both increased wear and heat, and clogging of filters, small apertures, and delicate equipment. Thus, equipment having moving parts requires particular care when designing for sand and dust protection. Sand and dust will abrade optical surfaces, either by impact when being carried by air, or by physical abrasion when the surfaces are improperly wiped during cleaning. Dust accumulations have an affinity for moisture and, when combined, may lead to corrosion or the growth of fungus.

In the relatively dry regions, such as deserts, fine particles of dust and sand readily are agitated into suspension in the air, sometimes reaching heights of several thousand feet, and lasting for hours. Thus, even though there is virtually no wind present, the speeds of vehicles or vehicle-transport equipment through these dust clouds can also cause surface abrasion by impact.

Although dust commonly is considered to be fine, dry, particles of earth, it also may include minute particles of metals, combustion products, and solid chemical contaminants. These other forms of dust may provide direct corrosion or fungal effects on equipment, especially if they are alkaline, acidic, or microbiological.

Because most equipment requires air circulation for cooling, removing moisture, or simply functioning, the question is not whether to allow dust to enter, but rather, how much or what size dust can be tolerated. The problem becomes one of filtering the air to remove dust particles above a specific nominal size. The nature of filters, however, is such that for a given working filter area, as the ability of the filter to stop increasingly smaller dust particles is increased, the flow of air or other fluid through the filter is decreased. Therefore, the filter surface area either must be increased, the flow of fluid through the filter decreased, or the allowable particle sized increased (i.e., invariably, there must be a compromise). Interestingly enough, a study by Pavia showed that, for aircraft engines, the amount of wear was proportional to the weight of ingested dust, but that the wear produced by 100 m dust is approximately half that caused by 15 m dust. The 15 m dust was the most destructive of all sizes tried.

Sand and dust protection, must be planned in conjunction with protective measures against other environmental factors. For example, it is not practical to specify a protective coating against moisture if sand and dust will be present, unless the coating is carefully chosen to resist abrasion and erosion.

Chapter 7

LONG-TERM NON-OPERATING RELIABILITY OF SELECTED ELECTRONIC PRODUCTS

Long-term non-operating reliability depends on the construction (materials and architecture) of the electronic products and assemblies. This chapter focuses on long-term storage effects of some specific electrical products. Parameters of concern are presented along with available field data.

7.1 MICROELECTRONIC COMPONENTS

In general, the circuitry within hermetic packages should not fail from moisture related storage conditions, although failures can arise and will be decreased in the following sections of this chapter. Typical failure mechanisms, corresponding sites in plastic packages, and failure modes are summarized in Table 7.1.

7.1.1 Die and Substrate Failures

In a typical daily environmental temperature cycle, the die, die attach, substrate, and package experience temperature differences and temperature gradient differences. Because these elements have different coefficients of thermal expansion, the bonds between these component parts can eventually fatigue over a long period of time.

As an example, the presence of edge voids in the die attach induces longitudinal stresses during daily environmental temperature cycling. Voids can act as microcracks that propagate during temperature cycling, resulting in weak adhesion, die lifting, and debonding of the die from the substrate

Table 7.1 Failure Sites, Failure Modes, Failure Mechanisms and Environmental Loads for a Plastic-Encapsulated Microelectronic Product.

Failure site	Failure mode	Failure mechanism	Environmental load	Critical interactions and remarks
Die edge, corner or surface scratch due to machining, dicing, or handling	Spalling crack, vertical crack, horizontal crack, electrical open	Crack initiation, crack propagation	Temperature gradients and changes	Encapsulant shrinkage, modulus of elasticity of encapsulant, CTE mismatch among die, die-attach and encapsulant
Die passivation defect, stress concentration on the passivation from bearing of sharp edge of filler particle	Transistor instability; corrosion of metallization, electrical open, shift in parametrics	Overstress, fracture, oxidation, electrochemical reaction	Cyclic temperature, temperature below glass transition temperature, humidity	Encapsulant shrinkage, sharp edges of filler, mismatch in CTE of chip, passivation and encapsulant
Die-attach void, crack, contamination site	Delamination of die, nonuniform transfer of stresses to the die pad from the die; eventual loss of electrical function	Crack initiation and propagation	Cyclic temperature	Moisture in die attach [Harada et al. 1992]; viscosity of die attach, die-attach thickness, CTE mismatch between die pad and die
Bonding wire	Breakage	axial fatigue	Cyclic temperature	Filler particle bearing, mismatch in CTE of wire and encapsulant

Failure site	Failure mode	Failure mechanism	Environmental load	Critical interactions and remarks
Ball bond	Electrical open, increased junction resistance, shift in electrical parametrics; bond lift, cratering, intermittence	Shear fatigue, axial overstress, Kirkendall voiding, corrosion, diffusion and interdiffusion, intermetallics at the base of the bond, neck in the wire	Absolute temperature, humidity, contamination	Lack of interdiffusion barrier layer
Stitch-bond heel, base, neck	Electrical open, increased junction resistance, shift in electrical parametrics, bond lift, cratering, intermittence	Fatigue, Kirkendall voiding, corrosion, diffusion and interdiffusion	Absolute temperature, humidity, contamination	Lack of interdiffusion barrier layer
Bond pad	Substrate cracking, bond-pad lift-off, eventual loss of electrical function, shift in electrical parametrics, corrosion	Overstress, corrosion	Humidity	CTE mismatch between bond pad and substrate, lack of passivation layer
Die and encapsulant interface	Eventual electrical open	Deadhesion, delamination	Humidity, contamination, temperature cycling about glass transition temperature	Residual stresses, formation of moisture layer at the interface, loss of adhesion, CTE mismatch
Passivation and encapsulant interface	Corrosion of metallization, shift in electrical parameters		Humidity, contamination, temperature cycling about glass transition temperature of encapsulant	Lead design, residual stresses, loss of adhesion, CTE mismatch between encapsulant and lead, lead pitch

Failure site	Failure mode	Failure mechanism	Environmental load	Critical interactions and remarks
Bond wire and encapsulant interface	Corrosion of die bond pad, increase in junction resistance, electrical open	Deadhesion, shear fatigue	Humidity, contamination, temperature cycling about glass transition temperature of encapsulant	CTE mismatch between encapsulant and wire
Encapsulant	Eventual loss of electrical function, loss of mechanical integrity	Thermal fatigue cracking, depolymerization	Temperature cycling about glass transition temperature of encapsulant	Delamination of die pad and encapsulant, corner radius
Leads	Reduced solderability, increased electrical resistance	Dewetting	Contamination, solder temperature	Lead finish porosity

or the substrate from the case. Depending on the die attach material, voids can form from melting anomalies associated with oxides or organic films on bonding surfaces, outgassing of the die attach, trapped air in the bonds, and shrinkage of solder during solidification. Insufficient plating, lack of cleaning, or even diffusion of oxidation-prone elements from an underlying layer, can generate voids during the melting of die attaches. In other instances, solder dewetting results in excessive voiding, especially when a solderable surface, a poorly solderable underlying metal, or excess soldering time produces an intermetallic compound not readily wet by the solder. Even under ideal production conditions, voids are often present due to solvent evaporation or normal outgassing during the cooling of organic adhesives. Although voids can stem from a number of sources, they are normally limited to an acceptable level through process control and rarely present a serious long-term reliability problem.

7.1.2 Stress-Driven Diffusive Voiding

Aluminum-silicon conductor lines in integrated circuits have a tendency to fail through the nucleation and growth of voids. Voids assume two dominant morphologies: a wedge-shaped form, and a narrow crack-shaped form. Conductor lines are typically passivated with a glass layer at approximately 700K and cooled to room temperature. The thermal coefficient of expansion of glass is an order of magnitude smaller than that of the conductor, so the cooling process imparts a significant tensile stress to the conductor line. This tensile field is not uniform, and stress gradients therefore cause diffusive mass transports. The passivated conductor lines are in a state of mechanical constraint, which prevents large-scale grain boundary sliding. With extensive mass transport and insufficient grain boundary sliding, void growth is inevitable.

The time to failure decreases with the decrease in conductor line width. This behavior is a direct consequence of higher stress states that develop in narrower lines. Calculations also show an increasing growth rate as the temperature increases, due to the diffusion coefficient, which increases exponentially with temperature, and the average far-field stress, which decreases quadratically with temperature [Yost, 1988].

In its present form, the model does not include the effect of temperature-dependent stress or stress relaxation that would occur as a highly stressed conductor is aged. At least two types of stress relaxation can occur. A homogeneous, or mean, field relaxation can take place throughout the entire conductor line and passivation layer, or a local stress relaxation process can occur as a result of the void process itself. The extent of the latter process depends on the metal grain size and the geometry of the debond region. Including stress relaxation effects would reduce the stress and significantly increase the time to failure at high temperatures. The rate of growth for crack-like voids is slightly faster than for wedge-shaped voids. The predicted time to failure for crack growth in 3 μm conductor lines is 1.5 years at 27°C, while the time to failure for wedge-like growth in 3 μm conductor lines is 2.3 years at 27°C. These results are consistent with industry experience [Yost, 1988]. Nevertheless, constraint cavitation remains a source of confusion and controversy. Confusion arises from the fact that most of the stress results from the deposition process, rather than from the difference in thermal expansion between the film and the substrate. High compressive nitride or oxynitride further contributes to the problem, because the excess hydrogen in silicon nitride embrittles aluminum, encouraging crack and void formation. To reduce cracks and voids, a diffusion barrier between the dielectric and the metal, or higher quality metals, can be utilized [Flinn, 1990].

More recent work on stress-driven diffusive voiding has revealed [Okabayashi, 1991] a more complex dependence lifetime on temperature than can be represented by a simple Arrhenius equation, as characterized by Yost [1988, 1989]. The Arrhenius formulation given by Yost can be explained by the stress relaxation in the conductor, with an increase in temperature which has not been accounted for. Diffusion, which is exponentially dependent on temperature, has been considered the rate-determining step in the Yost formulation. Okabayashi [1991] gave an analytical model for open-circuit failure for both wedge and slit-shaped voids and accounted for the stress relaxation occurring during void growth.

7.1.3 Wire Failures

The wires used to connect die bond pads to leads (or die bond pads to each other, in the case of hybrid packages) can fail from repeated flexing due to cyclic temperature changes. The most prevalent failure site is the heel of the wire. Gaffeny [1968], Villela [1970], Ravi [1972], and Phillips [1974] conducted extensive studies on wirebond failures and found they were due to the differential in the coefficients of thermal expansion of the wire and the package as the product was heated and cooled during temperature cycling. Failures in small-diameter wires may be inhibited by high loops, though the small dimensions of present-day packages impose stringent requirements on loop dimensions, making this solution largely impractical. Such a solution is only helpful for large-diameter wires, because wire stiffness restricts wire flexing. A cycle-to-failure estimate can be found in Pecht, [1989].

While temperature cycling can contribute to wire failure, the wires commonly used — gold for most plastic-encapsulated packages and aluminum for cavity packages and products that require high current density — are ductile enough to absorb large deformations.

7.1.4 Intermetallic Formation

Temperature cycling promotes intermetallic brittleness and growth. Bonded together, gold and aluminum present a problem when subjected to temperature cycling. Gold-aluminum intermetallics are more susceptible to flexure damage than pure gold and aluminum wires. Kirkendall voids occur at the interface between the wire and the bond pad when the contacting materials are gold and aluminum. Kirkendall voids form when either the aluminum or gold diffuses out of a region faster than the other can diffuse from the opposite side of that region. Vacancies pile up and condense to form voids, normally on the gold-rich side of the gold-to-aluminum interface.

Rates of diffusion vary with temperature and are dependent upon adjacent phases, as well as on the number of vacancies in the original metals. The compounds formed in the intermetallic are often called purple plague, because of the color of the gold-aluminum compound. Formed during thermocompression or thermosonic bonding, five intermetallic compounds can arise: Au_5Al_2, Au_2Al, $AuAl_2$, AuAl, and Au_4Al. Prolonged exposure to high temperatures, such as those encountered in tropical and desert environments, results in continued diffusion until the gold or aluminum is consumed. Figure 7.1 shows that the formation of significant amounts of purple plague can take years at 150°C and below; however, twenty years in storage at lower temperatures can be sufficient for the formation of noticeable amounts of purple plague.

Intermetallic compounds are mechanically strong, brittle, and electrically conductive. Temperature cycling can cause failure as a result of differential thermal expansion between the intermetallic and the surrounding metal, and can reduce mechanical bond strength due to voiding of the surrounding metal, which usually accompanies intermetallic formation. Gold-aluminum intermetallics are stronger than pure metals, and their growth is accelerated by increased temperature and temperature cycling [Philosky, 1973].

Newsome [1976] investigated on the effect of temperature on intermetallic formation in two thick-film systems (Owens-Illinois and EMCA 212B) and in conventional nichrome, nickel, and gold thin-film systems, conditioning them at 25, 75, 125, 150, 160, 175, and 200°C for 168 hours in an air-circulation oven. Table 7.2 lists observations of the various phases. The process of intermetallic growth was accompanied by volumetric expansion of the intermetallics, which are formed at temperatures in the neighborhood of 160 to 175°C, creating severe mechanical stress cracks at the vertical interface with the gold. Kirkendall voiding and mechanical stress due to volumetric expansion caused by intermetallic formation was found to increase in the presence of a larger volume of gold under the bonds. Philosky [1971] listed nine different rate constants for five gold-aluminum compounds; Gerling [1984] published an

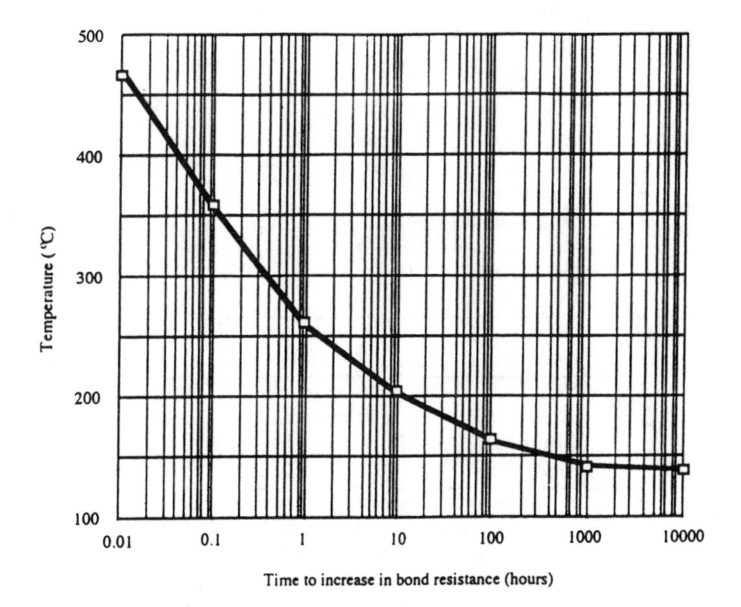

Figure 7.1 The formation of significant amounts of purple plague to cause a resistance change which will result in parameter drift [DM data, 1990]

Table 7.2 Summary of Optical and SEM Findings for the Three Gold Systems [Newsome, 1976].

		EMCA 212B	Owens-Illinois 99	Honeywell thin-film
Physical characteristics		Rough, pitted gold surface; irregular interface between gold and aluminum causes non-uniform intermetallic growth	Rough gold surface, but not as pitted as EMCA 212B; intermetallic growth is more uniform than for EMCA 212B	Uniform surface; gold 1/3 the thickness of thick-film conductors; at the interface between gold and aluminum, intermetallic growth is uniform
Constituents and energy dispersive spectrometry (EDS)		Thick-film ink with frit for mechanical adhesion to substrate; the frit contains silicon, bismuth, and cadmium as major constituents; the concentration of bismuth was found to be higher at the intermetallic diffusion front than in either the intermetallic or the gold Au_4Al, Au_5Al_2, AuAl, and $AuAl_2$ intermetallic phases were identified, with the AuAl and $AuAl_2$ not detectable for low temperatures;	Fritless thick-film ink with copper additive for wetting the substrate; no other detectable impurities the concentration of copper follows the same pattern as the bismuth in the EMCA 212B; Au_5Al_2, AuAl, and $AuAl_2$ intermetallic phases present, but no Au_4Al; AuAl and $AuAl_2$ appear at higher temperatures	Gold over nickel and nichrome; no detectable impurities Au_5Al2, AuAl, and $AuAl_2$ intermetallic phases present in detectable quantities, with AuAl and $AuAl_2$ appearing at the higher temperatures
Temperature	25°C	No detectable intermetallics	No detectable intermetallics	No detectable intermetallics
	75°C	Significant thickness of gold-rich intermetallic; visible Au_4Al (tan)	No detectable intermetallics	No detectable intermetallics
	125°C	Visible lateral voiding and annular cracks at gold interface; visible elevation of bond above surface of gold due to volumetric expansion of intermetallic; second gold-rich phase appears: Au_5Al_2 (tan)	First gold-rich intermetallic phase appears; identified as same phase that appears at this temperature for the EMCA 212B	First gold-rich intermetallic phase appears; identified as same phase which appears at this temperature for the EMCA 212B

extensive compilation of reported activation energies for various types of gold-aluminum-rich couples. Aluminum-rich gold-aluminum compounds have a high melting point and are relatively stable.

Newsome [1976] observed Kirkendall voiding in the neighborhood of 150 to 200°C (150°C for the EMCA 212B and Honeywell thin-film systems, 175 to 200°C for the Owens-Illinois system). Lateral voiding occurred at the aluminum interface at around 200°C. Kirkendall voiding in present-day packages is an impurity- or corrosion-driven reaction, temperature-independent below temperatures of around 150°C [Newsome, 1976].

However, in the presence of halogenated species, this process is accelerated at lower temperatures, around 125°C [Khan, 1986]. As these temperatures are far from those maintained in storage, Kirkendall voiding will only be a concern in corrosion- and impurity-driven reactions independent of temperature.

Gold wires bonded to bare copper leadframes and to copper thick-films in hybrids react forming three intermetallic phases: Cu_3Au, AuCu, and Au_3Cu. Temperatures above 300°C accelerate the rate of intermetallic compound formation. The time to decrease bond strength decreases with an increase in temperature [Hall, 1975]. Temperature time studies on thermocompression leadframe indicate a decrease in strength as a result of void formation. Pinnel [1972], Feinstein [1979a, 1979b], and Pitt [1982] studied gold thermosonic bonds to thick-film copper and found little strength degradation at 150°C for up to 3000 hours, and no failures at 250°C for over 3000 hours. The difference in hardness and material properties, including the coefficient of thermal expansion, between the intermetallic and the surrounding metal make the interface a potential failure site.

However, since such high temperatures are not encountered by components stored under natural environmental conditions, no intermetallic phases form.

7.1.5 Lead Seal Failure

In hermetic packages, most lead-seal failures could be tracked to defects that had escaped inspection or were a result of mishandling [Neff, 1986]. Microscopic examination of several hundred failed samples revealed that the defect or damage criteria could be classified into nine categories, including handling, radial cracks, poor pin-glass bonds, meniscus chips, eccentricity, poor glass-header bonds, weld problems, inadequate glass, and poor package quality. The relative percentages of failures attributed to each cause are documented in Table 7.3. Defect-induced and mishandling-induced lead-seal failures are thus a function of defect magnitudes, and have no dependence on temperature. These defects have a tendency to show up during storage, leading to failure of the lead seals.

7.1.6 Lead Corrosion

Leads undergo corrosion in the presence of moisture and ionic contaminants, which can result in a change in the electrical properties of the lead and loss of strength at the lead, leading to mechanical failure. Corrosion is usually localized at pinholes in the lead plating, diffusion sites of lead base metal to the lead surface, or plating cracks, all of which expose the base metal to the external environment. Cracks can appear in the lead plating, or even in the lead base material, during lead forming.

Table 7.3 Major Causes of Lead-Seal Failure [Neff, 1986].

Defects and failure modes	Percent of packages exhibiting failure mode
Handling damage: Most severe damage in lead seals has been attributed to poor handling in the form of bent leads, broken glass, and scratched gold plate.	95
Radial cracks: The cracks were generated in the seals starting from the pins outward, 20 to 60% of the distance to the outer diameter of the glass bead. Cracks resulted from side loading of pins due to improper fixturing, and from lead forming.	75
Poor pin-glass bonds: Meniscus seal often masks poor pin-glass bonds at the pin-glass junction. The defective packages pass incoming inspection and fail early in product life during lead forming.	60
Meniscus chips: Meniscus seal is the bond of the glass to the pin, which has flowed or wetted up to the pin to form a very thin circumferential glass-to-metal bond.	55
Eccentricity: Eccentrically located rectangular pins cause leaks. This type of structural defect seldom withstands thermal shock, such as a solder dip, which forces the sharp corner of the pin against a very small quantity of glass to produce a leak.	20
Poor glass-header bonds: Poor bonding at the outside diameter of the glass bead results in leaks early in the operational life of the product.	10
Inadequate glass: Glass seals fabricated with inadequate glass develop leaks in the form of radial cracks.	5
Poor quality package: Package defects incurred in manufacturing include lack of concentricity with rectangular pins, broken meniscus, broken plating, and glass broken away from pins.	60

Typically, thin nickel plating can cause lead cracking during bending or forming. Panousis [1976] examined various leads with nickel coatings varying in thickness from 0 to 41 μm, and bent through 90°. A crack was defined as a discontinuity on the lead surface exceeding 10% of the lead thickness. It was found that leads did not crack if the nickel plating thickness was less than 1.0 μm; a nickel thickness of about 2.7 μm resulted in cracks with depths less than 10% of lead thickness. A nickel thickness of greater than 5.7 μm resulted in cracks larger than 10% of the lead thickness. Zakraysek [1981] evaluated various lead coatings on bent leads with respect to corrosion resistance in salt-spray tests. He found that, when an electroless nickel undercoat was used, tin and tin-lead finish parts generally passed, while the gold finish parts all failed, because the cracks induced during the bending of the leads exposed the base material to corrosion. The presence of gold accelerated corrosion of the base metal. Tin plate over electrolytic nickel was found to be superior to tin over copper, electroless nickel, or bare Kovar after twenty-four hours of exposure to a salt-fog atmosphere. Tin over electroless nickel performs poorly because highly stressed electroless nickel deposits can develop cracks and expose the base metal to the corrosive environment. Neither 99.7 nor 99.9% gold plate passed the corrosion test when electroless nickel was employed as the undercoat.

Although leads can undergo corrosion anywhere on their surfaces, they are particularly susceptible to corrosion at the interface between the glass seal of the leads.

Since the lead coating is applied after the lead sealing, the part of the lead covered by the sealing glass is not coated. This makes it highly prone to corrosion in the presence of moisture and ionic contaminants. If the lead seal does not form a bond at the lead surface, it will provide a pathway for moisture to contact the bare lead surface. During assembly and handling the lead is often bent, causing the glass meniscus to crack at the glass-lead interface and exposing the bare lead surface underneath to the external environment.

7.2 MILLIMETER WAVE COMPONENTS

Millimeter wave components include Gunn diodes, varactor diodes, mixer diodes, monolithic integrated circuit two-terminal products, MIMIC three-terminal products, circulators, and intermediate frequency amplifiers. In generally, after long-term inactive storage, millimeter wave components experience little or no aging effects when hermetically sealed. The interested reader is referred to the work of Melmed [1991], where much of this section was taken from.

Moisture can cause corrosion of integrated circuit metallization paths (open circuits) or corrosion of other metal parts which can cause component/assembly failure. The reader is referred to Chapter 6 for more details. Also, foreign matter, such as dirt or debris bridging metallization paths, can cause short circuits.

7.2.1 Gunn Diodes

Gunn diodes are active devices used as radio frequency (RF) sources in MMW radars. Parameters of concern in long-term inactive storage include frequency drift, power output, linear tuneable bandwidth corrosion of ribbon bonds, and threshold current. A key parameter which reveals significant degradation is the threshold current. Small threshold current changes are attributable to measurement errors and changes of ±10% or more indicate degradation. Power output and operating frequency become noticeable when the threshold current of a Gunn diode change by ±10% or more. Since Gunn diodes are of non-planar designed, material migration is out of usually susceptible. Concerning ion migration, the Gunn diode impurity doping level is too low to be susceptible to ion migration.

7.2.2 Varactor Diodes

The varactor diode is a variable capacitance semiconductor element (also called a voltage-controlled capacitor). Its capacitance varies with the applied negative bias voltage. Compared to the normal semiconductor diode, the capacitance of the P-N junction in a varactor diode is emphasized instead of minimized. Varactor diodes are made of GaAs for most MMW radar applications, in order to keep insertion losses at a minimum. It is commonly referred to as a tuning diode and has the function of allowing the radar frequency to be changed or tuned with time in a controlled fashion. This process is called frequency modulation.

Parameters of concern in long-term inactive storage include linearity of capacitance change with applied voltage, magnitude, and insertion loss.

7.2.3 Mixer Diodes

Mixer diodes provide frequency conversion, whereby an RF signal is combined with a local oscillator signal in a nonlinear fashion resulting in a desired lower difference frequency called the intermediate frequency (IF). Mixer diodes used in millimeter wave radar applications are usually of Schottky barrier construction. Schottky barrier diodes are made of either silicon or GaAs, which are preferred for the MMW frequencies.

Parameters of concern in long-term inactive storage are conversion loss, noise temperature ratio, insertion loss, noise figure, change in nonlinear resistance characteristics, and forward voltage drop. Schottky barrier mixer diodes with bonded connections and hermetically sealed in metal-ceramic packages were tested in 1974. The parameter chosen to reveal any significant aging effects was the forward voltage drop. The forward voltage drop is the difference in voltage between two points due to the loss of electrical pressure as a current flows through an impedance. This test was significant because the expected degradation from long-term storage is the diffusion of gold through the Schottky barrier and into the GaAs, thus changing the electrical characteristics of the diodes. Diode suppliers are aware of this problem and add a platinum layer, at least 300 angstroms thick, to act as a diffusion barrier. The measured variations between 1974 and 1987 are well within normal measurement accuracy and there is no indication of any degradation.

7.2.4 Monolithic Integrated Circuit Two-Terminal Devices

Monolithic integrated circuit two-terminal devices are monolithic GaAs devices with a number of the conventional two terminal diodes on a single GaAs substrate. This construction eliminates assembly requirements associated with hybrid microwave circuits.

Parameters of concern in long-term inactive storage are RF conversion loss in a mixer chip mounted in fixture, and R_s while in a fixture, where R_s = series resistance of the diode.

7.2.5 Monolithic Integrated Circuit Three-Terminal Devices

MIMIC three-terminal devices are used as RF oscillator and mixer circuits. They are constructed from GaAs field effect transistors which are commonly referred to as three-terminal devices and are integrated on a single GaAs substrate.

Parameters of concern in long-term inactive storage are changes in metal semiconductor field effect transistor (MESFET), gate length (0.25 micron and 1 micron) HEMT, transconductance gm - figure of merit, cutoff frequency f_c, and response of the device to bias requirements.

7.2.6 Circulators

Circulators are multiport RF devices which allow energy to enter the input port and exit only from a given output port. In principle, the ferrite circulator can offer separation of the transmitter and receiver without the need of the conventional duplexer configurations. It is often used to isolate a transmitter and a receiver when both are connected to the same antenna.

Parameters of concern in long-term inactive storage are isolation between the transmitter and receiver, impedance match between the circulator and antenna, and effects on magnetic moments.

7.2.7 Intermediate Frequency Amplifiers

There are two common types of intermediate frequency (IF) amplifiers: hybrid IF amplifiers and monolithic IF amplifiers. Hybrid construction usually refers to thick or thin film technology in which individual component chips are mounted on a circuit board and connected by wire bonds. The function of hybrid IF amplifiers is to take low-level IF signals and amplify them to a level which can easily be processed by down stream signal processing electronics. Although having the same function as the hybrid IF amplifier described above, the monolithic IF amplifiers are constructed on a single substrate. The monolithic configuration is used when the state-of-the-art technology will allow the level of integration.

The key parameter of concern in long-term inactive storage is noise that could affect system performance for both amplifiers.

7.3 INFRARED COMPONENTS

Some important infrared components include lead selenide detectors, coolers, and optics. Infrared components may experience some effects of chipping and flaking or mirror and lens coatings and possible lead selenide detector damage if exposed to moisture during long-term storage. Hermetic seals will prevent corrosion of metal parts. While some component failures can also be traced to multi-environmental long-term storage, the sensitivity of some detectors increases with storage time.

7.3.1 Lead Selenide Detectors

Infrared technology has been employed to sensors to detect the thermal contrast of targets against the natural thermal energy of the earth. Thermal contrast exists in nature, since all bodies above absolute zero emit infrared energy. The energy emitted is a function of the temperature of the body and its emissivity. Intrinsic photoconductive detectors include silicon, germanium, lead sulfide, lead selenide, indium arsenide, and indium antimonide. When considered as a circuit element, a photoconductive detector behaves much like a variable resistor. The detector is connected in series with a load resistor and a bias battery. Photon-induced changes in the conductivity of the detector modulate the current flowing through the detector and load resistor. The signal is taken across the load resistor and capacitively coupled into a preamplifier. Lead selenide detector arrays can be packaged in another configuration which involves transconductance preamplifiers where capacitive coupling is not required.

Parameters of concern in long-term inactive storage are: peak wavelength which should remain within required bandwidth, responsivity, resistance, time constant, and crosstalk between adjacent elements.

One company tested their PbSe detectors in a 10-year period (1974-1984) on three separate lots. The test result indicated minimal changes in resistance, a maximum increase of 1.33 in signal and a maximum increase of 1.31 in D-star. The PbSe displays significant performance changes while in long-term (10-year) dormant storage.

A second company's long-term storage data on PbSe detectors indicate an average increase of 32% in responsibility and D-star over a 10-year period. The comparison of the high energy thresholds relative to the signal generated by an IR system before and after long-term storage indicates the target signal is closer to the high energy threshold after the 10-year storage. Overall, this company claims that long-term dormant storage has a positive impact on their sensor's performance. According to this company's claims, the change in detector performance will provide:

- Improved target detection under low signal-to-noise conditions
- Improved aim point resulting from increased IR detection
- No effect on detection in clutter limited scenarios
- No effect on decoy rejection
- Potential increase in false alarms under particular background conditions

A third company stored PbSe detectors for 6 years; 14 detectors have been tested for signal response, noise voltage, signal-to-noise ratio, and resistance. These detectors show remarkable stability considering they are not contained within any kind of package. The company claims that packaging in hermetically sealed containers assures long life and a high degree of stability even in very hostile environments.

7.3.2 Coolers

There are two kinds of coolers used in infrared detector packages: thermoelectric coolers and cryogenic coolers. Thermoelectric coolers are commonly made by Bismuth Telluride. The principle of the thermoelectric cooler, the Peltier effect, is used for cooling by simply reversing the current's direction. Temperatures differentials of 60 to 65°C can be achieved for a single couple, or stage, using an infrared detector as a heat load. If several couples or stages are cascaded, lower temperatures can be reached. Joule-Thomson coolers use a Joule-Thomson cryostat, a miniature gas liquefier that can be placed directly in the coolant chamber of a detector package. In its most common form, a cryostat consists of a cylindrical mandrel carrying a helically wound coil of finned metal tubing. The finned tubing serves as the countercurrent heat exchanger. The cryostat slides smugly into the inner steam of the dewar, and the steam of liquefied gas is directed toward the back of the surface carrying the sensitive element. After expansion, the cooled gas flows along the heat exchanger and extracts heat from the incoming gas. Although most cryostats are designed for use with nitrogen or argon, they can also be used with the various members of the Freon family to produce a wide range of temperatures.

Parameters of concern in long-term inactive storage for thermoelectric coolers are cooling capacity degradation and Z (figure of merit). Parameters of concern for cryogenic coolers are capacity degradation and gas leaks in the coolant bottle.

7.3.3 Optics

IR optical systems combine the advantages and disadvantages of both reflective and refractive optics. Refractive optics incorporate lenses while reflective optics incorporate mirrors. In general, the lens system consists of an objective, to gather and condense incoming IR radiation, and a field lens to focus radiation on the sensitive surface of the detector. Objective and field lenses may be either simple single lenses or complex multiple lens systems, depending upon the degree of resolution, the correction required for various aberrations, the focal length, and instantaneous field of view required of the optical system. Reflective optics use optical mirrors. Many smart munitions use folded reflecting or mirror systems to direct beams of radiation. The advantage of the folded reflecting system is that longer focal lengths can be obtained in a smaller space.

Parameters of concern in long-term inactive storage for optical lenses are changes in absorptive and reflective characteristics of the coating applied to the lens surface, and effects of changes in lens aberration and other optical qualities. Parameters of concern for optical mirrors are degradation of the polished mirror surface, and shift of the mirror in its mount.

7.3.4 Optical Coating

Antireflection coating and high-reflection coatings are used in IR optical systems to achieve high optical efficiency. Reflective or reflection losses occur when IR radiation is incident upon an air-material interface. A thin film of coating can be applied to the surface by vacuum evaporation in order to eliminate completely the reflection at a given wavelength. An antireflection coating for optical materials used in air must meet two criteria; its index of refraction must be equal to the square root of the index of the optical material to be coated, and its optical thickness must be equal to one-fourth of the wavelength at which minimum reflection is to occur. Optical thickness is the product of the index of refraction and the physical thickness of the coating. Optical thickness changes with the angle of incidence, however, the effect is negligible for angles less than 30°. Table 7.4 lists the materials that have been used for antireflection coatings. After the introduction of evaporating metal film techniques during World War II, metal films have been used in virtually all mirror high-reflection coatings instead of the traditional chemically deposited silver high-reflection coating method. The reflectances of most metals increase at longer wavelengths. The spectral reflectance of various evaporated metal films is shown in Table 7.5.

Parameters of concern in long-term inactive storage for optical coatings are chemical degradation, reflectance, chips, cracks, flaking, blemishing, and changes in optical characteristics.

Table 7.4 Materials for Low-Reflection Coatings [Melmed, 1991].

Material	Useful spectral range (μ)	Average index of refraction
Cryolite ($A1F_3N_aF$)	0.2 - 10	1.34
Magnesium fluoride (M_gF_2)	0.12 - 5	1.35
Thorium fluoride (ThF_4)	0.2 - 10	1.45
Cerium fluoride (C_eF_3)	0.3 - 5	1.62
Silicone monoxide (SiO)	0.4 - 8	1.45 - 1.90
Zirconium dioxide (ZrO_2)	0.3 - 7	2.10
Zinc sulfide (ZnS)	0.4 - 15	2.15
Cerium dioxide (CeO_2)	0.4 - 5	2.20
Titanium dioxide (TiO_2)	0.4 - 7	2.30 - 2.80

Table 7.5 Reflectance of Evaporated Metal Films [Melmed, 1991].

Wavelength (μ)	Reflectance (%)				
	Aluminum	Silver	Gold	Copper	Rhodium
0.5	90.4	97.7	47.7	60.0	77.4
1.0	93.2	98.9	98.2	98.5	85.0
3.0	97.3	98.9	98.3	98.7	94.5
5.0	97.7	98.9	98.3	98.7	94.5
8.0	98.0	98.9	98.4	98.7	95.2
10.0	98.1	98.9	98.4	98.8	96.0

7.3.5 Optical Filters

Optical filters are required in most IR systems in order to pass IR radiation of a particular wavelength band through the optical system to the sensitive surface of the detector. This process is known as spectral filtering. The properties of optical filters are very dependent upon the optical materials and methods of construction employed. The reflection or interference filter uses interference effects to reflect rather than to absorb unwanted radiant energy. Long-wavelength cutoff filters pass IR radiation only up to a particular wavelength. Short-wavelength cutoff filters block all IR radiation below a certain wavelength. Interference filters are made by the vacuum deposition of several layers of dielectric material onto a suitable substrate, precisely controlling the index of refraction and the thickness of each layer.

Parameter of concern in long-term inactive storage is degradation of optical qualities (change of spectral characteristics).

7.3.6 Focal Plane Header Assembly

The focal plane header assembly is an assembly which could contain various IR detector components, cooling mechanisms, amplifiers, impedance matching networks, and IR filters. It is configured in different ways depending upon system requirements. Four kinds of common focal plane header assembly configuration are (1) uncooled header assembly (preamp not included), (2) cooled header assembly (preamp not included), (3) uncooled or cooled header assembly (includes preamp), (4) hermetic seal or inert gas back-fill configuration.

Parameters of concern in long-term inactive storage are hermetic seal leak, corrosion of metal parts, thermal bond at junction, and mechanical soundness.

7.3.7 Infrared Transducer

The function of the infrared transducer is to detect the presence of targets which radiate energy in the infrared portion of the spectrum. An infrared transducer system is an assembly with optics (lens and mirror), filters, detectors, and coolers. The main purpose of optical systems in infrared transducers is to focus the desired rays of energy onto the IR detectors. The detectors are the components which actually sense the presence of the target emitted energy and transform this into a recognizable electrical signal. Detectors are often cooled by means of thermoelectric coolers or cryostats. The electrical output signal of the detectors is fed to a preamp which amplifies and filters this signal.

Parameters of concern in long-term inactive storage are hermetic seal problems, possible degradation of thermal bond at junction, metal parts might rust, degradation of cementing compound, and parameters from the combination of its parts (given previously).

7.4 PRINTED CIRCUIT BOARDS AND SOLDER JOINTS

There are no specific recommendations regarding how long a printed board can be stored, or specific recommendations on temperature and humidity requirements for optimum storage conditions. Nevertheless, general guide-lines information is available, along with some test data on printed board storage and test methods for printed board solderability.

Requirements and test methods for solderability are specified in J-STD-003, "Solderability Test for Printed Boards," which contains four tests with established accept/reject criteria: the edge dip test, the rotary dip test, the solder flow test, and the wave solder test. In addition, it contains a wetting balance test which currently does not have established accept/reject criteria.

In order to maintain solderability, it is crucial that proper handling and storage procedures for printed boards are followed. Although the IPC does not have much documentation on printed boards storage, some guidelines information is contained in Section 7.4 of IPC-AJ-820, the "Assembly-Joining Handbook." Section 7.4 contains general information on packaging, including bags and Craft paper, along with references to military specifications for the packaging of printed boards, components, and hardware. IPC-AJ-820 does not contain information on desiccants, industry suppliers should be contacted.

In addition to the aforementioned two documents, the IPC does publish some data on solderability evaluations of printed boards. IPC-TR-462, "Solderability Evaluation of Printed Boards with Protective Coatings Over Long Term Storage," discusses the results of a round robin study on the effect of coating used for preservation of printed board solderability. The coatings were evaluated over a two-year period and include tin-lead plated and solder coated boards, along with copper-only boards that were protected by organic coatings. The general conclusions of the report state that both the tin-lead plated and solder-coated boards exhibited good soldering performance after two years of storage, but the copper-only boards did not solder as well. The report also states that solderability acceptance criteria for printed boards should not be based on coating types or thicknesses, but on functional testing as described in the IPC solderability documents (J-STD-003).

When components are mounted to a printed wiring board assembly, thermal stresses can arise due to coefficient of thermal expansion mismatches between the materials. The time to failure will depend on the materials and geometry, the strain amplitude, the maximum and minimum cycling temperatures, and the frequency of thermal cycling [Agarwala, 1985]. Engelmaier [1989] provides simplified equation for leadless surface-mounted components.

7.5 CABLES AND BATTERIES

Cables and batteries, like other general electronic components, show similar deleterious effects when exposed to moisture. Potting and hermetic seals may be required to prevent corrosion of metal parts.

7.5.1 Cables

The basic structure of the cable circuits is rolled annealed copper encapsulated in polyimide film with exposed tin-lead solder pads. Flex circuits should be shipped in individual sealed bags with a desiccant and humidity indicator to assure the shelf life of the solder pads. One company, which built flexible circuits in 1964 for Minuteman missiles, claim that if a circuit has been correctly designed and built for an application and environment, the specifications they test to today will insure its operation over the life of the assembly. To date they have never received any report of flex circuit failures due to aging.

7.5.2 Batteries

Lithium aluminum/FeS_2 thermal batteries were tested after being stored for 6 months and 12 months at 160°F. Comparisons of the test data before and after the storage shows no degradation had occurred. In fact, the average battery life under the high test temperature increased by about 15 seconds after 6 and 12 months of storage. The peak battery voltage of the cold test temperature batteries were lower by about 0.5 V after a 12-month storage. The activation time showed a spurious rise of approximately 30 ms at the 6-month period, but returned to original values at the 12-month test. The testing was performed by an independent laboratory and the activation time data collection method was changed over this time period. In conclusion, the company claimed that thermal batteries are inherently capable of withstanding 15 to 20 years of shelf life. This capability is due mainly to the solid, nonconductive nature of the electrolyte salts at room temperature, the control of moisture during construction, and the hermetic seal of the battery.

Chapter 8

TESTING AND MAINTENANCE

In order to successfully test and maintain a product subjected to long-term non-operating condition, it is important to understand the design, the manufacturing process, the loads throughout the life cycle, the potential failure mechanisms, and the cost penalties associated with each design and manufacturing decision. As discussed in previous sections, the term loads is used in a generic sense and includes mechanical, thermal, electrical, radiation, and chemical stimuli that can affect performance during non-operating conditions.

8.1 TESTING

Most electronic products are likely to demonstrate high reliability when subjected to non-operating conditions. Evaluating the distribution of failures and life expectancy can therefore be difficult because of the very long test periods required to obtain sufficient failure data. For this reason, overstresses and accelerated wear-out stresses are often employed. Overstress tests can be conducted when qualifying a product for potential overstress non-operating loads. Testing for wear-out failure mechanisms is usually accomplished by accelerated testing which allows for compression of test time.

Accelerated testing involves measuring the reliability characteristics of the product quantitatively under more severe stress conditions than the normal operating level, in order to induce wear-out failures within a reduced time period. The advantages of accelerated life tests are both economic savings and quick turnaround during the development of new products or of mature products subjected to manufacturing and workmanship changes. The severity of the applied stress in accelerated wear-out testing is usually selected based on achieving a reasonable test time compression without altering the fundamental failure mechanism. The results from the tests are then extrapolated using a quantitative acceleration transform to give a lower bound estimation of life for the product.

In accelerated test design and test data interpretation, it is necessary to understand which type of stress accelerates a particular failure mechanism and how varying the rate of application or magnitude of the stress influences the life of the product. The technique of accelerated testing includes

- Selecting the appropriate stress parameters to be accelerated;
- Determining the magnitude to which the stress parameter(s) should be accelerated;
- Designing the test procedure, such as step-stress acceleration; and
- Extrapolating the test data from physical failure models.

The degree of stress acceleration is usually controlled by an acceleration factor, defined as the ratio of the life under normal use conditions to that under the accelerated condition. The acceleration factor should be computed for each product and each field condition, from a quantitative transform that gives a functional relationship between the accelerated stress and the damage caused.

Some caution and discretion, as expressed in the following guidelines, must be exercised when using accelerated testing:

- The failure mechanism(s) tested in the accelerated environment must reflect the failure mechanism(s) which will arise during operation after some period of non-operation.
- The accelerated environment can be confidently extrapolated to some long-term non-operating conditions.
- The properties of the materials do not change under accelerated stress unless these changes are consistent with life-cycle experiences.
- The shape of the failure probability density function under normal and accelerated conditions is transformable.

To determine the limits of the stress that can be applied to a product without causing atypical failures, a step stress test is often conducted. A progressively increasing stress is applied to the product for a constant time interval. The number of failures that occur at each stress level are recorded, and the failure modes are investigated. The stress step at which products start failing with atypical failure modes sets the stress limit at which an accelerated test can be carried out. The step stress test enables the product to be tested at different stress levels using the same sample, which makes it fast and inexpensive. However, the estimate of the maximum allowable stress on the product is not very precise, because the test is based on the assumption that failures at a particular stress step are independent of the preceding steps. In practice, the effects of stresses are cumulative and the data obtained are only approximate. Nevertheless, it is a useful method of determining the stresses that can be used to conduct more accurate (and more time-consuming) stress tests.

8.2 TESTS TO ADDRESS NON-OPERATING CONDITIONS

8.2.1 Temperature Cycle

Temperature cycle testing can be conducted to assess the effects of thermal expansion mismatches among the different components and interfaces within a product. During temperature cycle testing, products are held at the cold dwell temperature long enough to establish thermal stabilization and appropriate creep and stress relaxation of the materials. Following this cold dwell, the products are heated to the hot dwell, where they remain for another minimum time period. The dwell at each extreme and the two transition times constitute one cycle. Test duration will vary with the product. Common failure modes include parametric shifts and catastrophic failures; common failure mechanisms include wirebond fatigue, cracked or lifted dies, and package failure.

8.2.2 Pressure-Temperature-Humidity

The pressure-temperature-humidity test accelerates the effects of moisture penetration (and corrosion) failure mechanisms. Conditions typically employed during this test are a temperature of 125°C or greater, pressure of 15 psig or greater, and humidity between 80 and 95%. The autoclave test is a special case, generally conducted at 121°C, 100% relative humidity, and pressure of 15 psig. These tests can be preceded by temperature cycling and followed by others to further accelerate the corrosion failure mechanism. Common failure modes include parametric shifts, high leakage, and mechanical failures. Common failure mechanisms include corrosion, contaminants within the package materials, and poor package sealing.

8.2.3 Mechanical Shock

This test examines the ability of the product to withstand a sudden change in mechanical stress, typically due to abrupt changes in motion during handling, transportation, or actual use. Common failure modes include opens, shorts, excessive leakage, and mechanical failure. Common failure mechanisms include fracture, fatigue, and wear.

8.2.4 Variable-Frequency Vibration

This test examines the ability of the product to withstand deterioration due to mechanical resonance. Common failure modes include opens, shorts, excessive leakage, and mechanical failure. Common failure mechanisms include fracture, fatigue, and wear.

The accelerated test under random stress is usually conducted by elevating the power spectral density (PSD) function of stress, displacement, or acceleration of the product. However, the value of the PSD function is not explicitly related to the fatigue life, as in the S-N curve. Only for some very special cases, such as narrow-band, zero-mean Gaussian stress processes, is there a relation between PSD and fatigue life. As fatigue damage is sensitive to the sequence of random stress amplitude, the possibility of accelerating the damage parameters in the frequency domain is questionable; it may be that no unique acceleration factor exists for this case. Moreover, for random fatigue accelerated tests, the signals can be displacement or acceleration only if the relationship between displacement and stress or between acceleration and stress is linear.

8.2.5 Integrity

This test examines the mechanical properties of a product's leads, welds, and seals. Various conditions can be employed: tensile loading, bending stresses, torque or twist, and peel stress. In addition, sand, dust, radiation, high temperature, and other stress conditions can be employed to assess material ageing, wear, abrasion, and other failure mechanisms arising from these environmental conditions. The failure is determined visually under 3X to 10X magnification.

8.3 MAINTENANCE

Maintenance is conducted to ensure that previously qualified product parameters are within specified tolerances through the monitoring, verification, and control of critical material and stress variables. Parameter variability due to long-term non-operating conditions may be due to any one or a combination of the following factors:

- Changes in material properties (e.g., due to ageing, creep)
- Changes in structure and interfaces (e.g., due to corrosion, intermetallics, fatigue, creep)
- Unintended stresses (e.g., contaminants, particles, lightning, vibration) in the environment

Product evaluation testing can be considered a maintenance task, in the sense of providing an audit to ensure that the product conforms to the control limits of the production processes. In some cases, a defective product may be eliminated immediately, preventing additional costs from accruing.

Whenever possible, maintenance should detect a defect without "stressing" the product. Examples include visual examination and functional testing. When using stress tests, the crucial step is to determine the appropriate magnitude or intensity of the stress required to precipitate the defect without reducing the useful life of the product.

Periodic maintenance (inspections and functional tests) of stored electronics can help to ensure their reliability and readiness for use, to detect any deterioration during storage, and to implement measures for impeding further deterioration. The inspection process involves:

- Visual inspection of the storage environment, storage containers like cardboard boxes, inner packing, and the microelectronics;
- Tests of the microelectronics for fatigue, cracks, and contamination that cannot be detected visually.

Products with a finite shelf life (which deteriorate during dormancy) are inspected before products of infinite shelf life (which do not deteriorate while dormant). To determine the intervals at which the products need to be tested and inspected, the inherent deterioration of the materials and the repair time, cost, and complexity of the products must be considered along with the storage environment and degree of packing. Improper storage methods and storage periods exceeding the shelf life increase the degree of deterioration.

Table 8.1 provides some design guidelines to make the maintenance process easier.

Table 8.1 Design for Maintenance.

1.	Design for minimum maintenance skills. Some technicians are neither well trained nor well motivated.
2.	Group parts and subsystems so they can be easily located and identified.
3.	Provide for visual inspection.
4.	Provide troubleshooting techniques and test points. Use plain marking and adequate spacing and accessibility.
5.	Label units. Labels on top of components, parts, and structures (or from the direction of normal human access) should agree with instruction manuals.
6.	Provide handles or hand-holds on heavy components for easy handling.
7.	Design for minimum tools. Special tools and laboratory test equipment may not always be available.
8.	Design for minimum adjusting. Adjustment for shift, drift, and degradation should not be necessary in most cases.

Failure analysis of defects detected during the maintenance process is the key to formulating effective corrective action to remove the root cause of the defect. A typical defect analysis should include

- Details of the defect or failure event,
- The circumstances leading to the defect or failure,
- Process irregularities,
- Mechanisms inducing defects or failures, and
- Recommendations for eliminating the defects.

Chapter 9

A FRAMEWORK FOR NON-OPERATING RELIABILITY ASSESSMENT

The electronics industry recognizes that understanding potential failure mechanisms leads to eliminating them cost-effectively. Consequently the preferred approach to reliability assessment utilizes knowledge of failure mechanisms to encourage robust design and manufacturing practices.

This chapter describes a framework for assessing the reliability of non-operating electronic products based on the discussions in previous chapters. The framework encourages the assessment of reliability during the design phase; facilitates the evaluation of new materials, structures, and technologies; and guides the user in developing tests and screens.

9.1 ELEMENTS OF THE FRAMEWORK

An effective framework for non-operating reliability assessment must be proactive and incorporate reliability assessments in the design process. The flowchart in Figure 9.1 illustrates how the reliability assessment process is performed in parallel with traditional electrical engineering processes and prior to manufacturing. The elements of the framework are discussed in this section.

9.1.1 Constraints

Constraints specify degrees of freedom, or opportunities, in product development. A key task of the design team is to assign values to the design variables within the bounds of the constraints. Design constraints constitute the fixed information given to the design team. Major design constraints for electronic products include parameters defined by the passive and active elements that need to be packaged, as well as the platform on which the product elements will be mounted; data on manufacturing, assembly, storage, transportation, usage, and repair environments; manufacturing limitations and capabilities; material selection; product architecture; production volume; yield; cost; and schedule.

Contractual requirements, company policy, and user preferences may also be part of constraint information. For example, some military contracts specify the use of hermetic components, some companies manufacture only a particular package style, and some users prefer surface-mounted components.

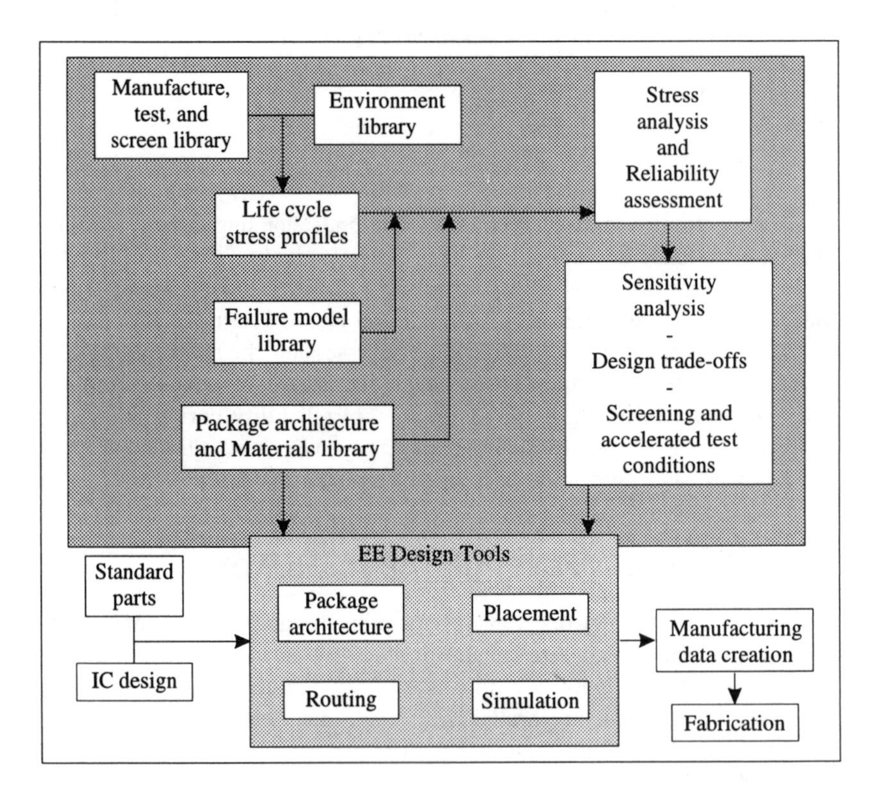

Figure 9.1 Reliability assessment tools must feed directly into the design phase of electronic product engineering

9.1.2 Materials

The materials library includes properties for all product elements, including components, interconnects, and mounting platform. These properties provide a basis for selecting materials for different product elements, and also constitute inputs for the physics-of-failure models used in reliability assessment.

9.1.3 Product Architecture

The product architecture is the specification of the geometry and materials of a product. Design aids are provided to help the user select technology, component types including dual in-line package (DIP), quad flat-pack (QFP), or pin-grid array (PGA) and mounting platform. The user can also specify the electrical parameters and physical dimensions of the active and passive elements to be packaged (Figure 9.2 shows some of the architecture selection paths).

9.1.4 Non-Operating Environment

The non-operating environment library includes complete descriptions of common storage/dormancy environments in terms of characteristic load parameters —temperature (minimum, maximum, and average), magnitude and number of temperature cycles per year, relative humidity (RH) (minimum, maximum, and average), RH cycle, vibration load (acceleration power, spectral density, frequency, waveform, and vibration mode), maximum acceleration load, weather and environmental conditions; and radiation.

The mission profile refers to the magnitude and duration of all the loads the product is subjected to during its life cycle. The mission profile allows the user to specify the environment, test, and screen durations to which the product will be exposed during its mission life. Mission profile information is used by the failure-model library to evaluate the dominant failure mechanisms in the product architecture under the specified loads.

9.1.5 Defects Database

The defects database addresses material flaws, manufacturing-induced defects, and assembly defects. Defects are classified in terms of their occurrence in fundamental constructs. Defect information includes thresholds, origins, strategies to reduce occurrence, accelerating stresses and environmental loads, physics-of-failure models characterizing defect magnitude and test or operational loads, and failure mechanisms actuated in the field and during testing.

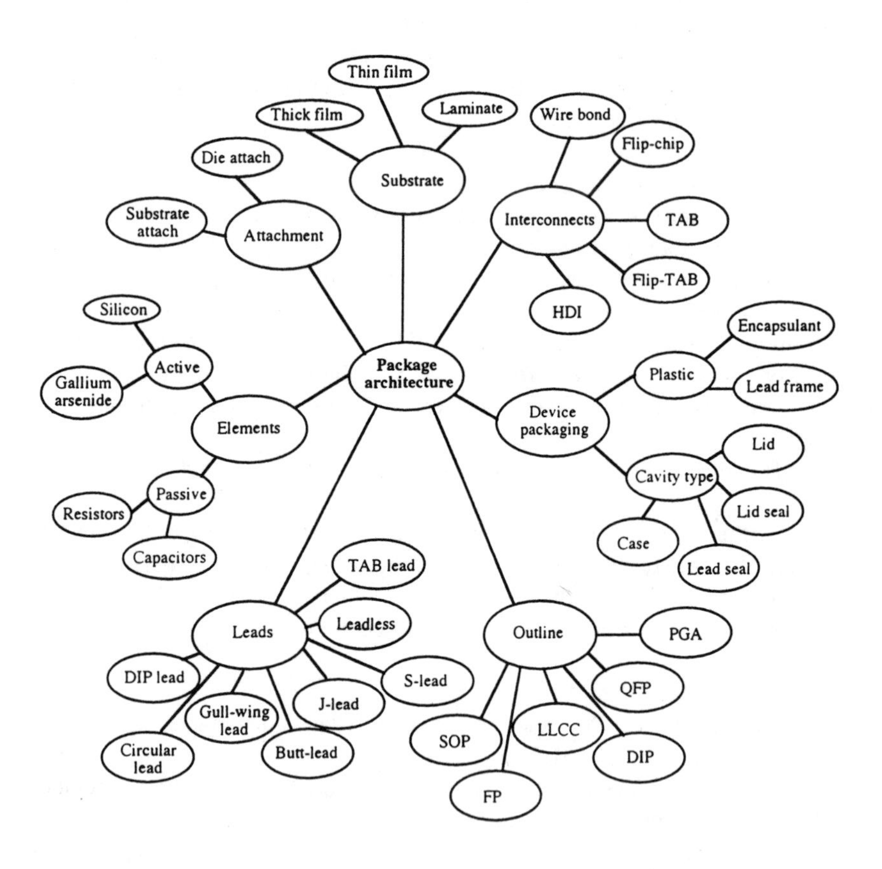

Figure 9.2 The product is described by specifying the dimensions and materials for each component and component element. A microelectronic package architecture breakdown is shown here

9.1.6 Reliability Assessment and Sensitivity Analysis

The reliability analysis estimates the time-to-failure for dominant failure mechanisms and ranks them accordingly. The failure models used to evaluate each mechanism can be selected from the failure mechanism library or can be user-specified, depending on the application.

Reliability information can be used to assess whether the package will survive for its designed-for life. If the time to failure for the mechanism with the lowest time is less than the desired mission life, then the sensitivity of the failure mechanism to design parameters can be evaluated until system reliability goals are met.

Cumulative sensitivity curves evaluate the product life for multiple stresses; the user can plot the product life vs percentage change from a nominal stress value. This tool allows the user to identify stress thresholds below which lowering stress magnitudes will produce no additional benefit in terms of added life. These thresholds can be identified as values of stresses for which the projected time to failure is well beyond the specified mission life of the module (Figure 9.3).

9.1.7 Tests and Screens

The test and screen library includes a complete description of tests and screens in terms of their characteristic stresses, which include temperature, relative humidity, pressure, electrostatic discharge, radiation, and refined forces (e.g., bond pull or die shear). The user can select the stresses that will be part of the testing or screening requirements. Each stress is characterized by a set of parameters, which include maximum, minimum, dwell at maximum and minimum, ramp time maximum-to-minimum, ramp time minimum-to-maximum (for temperature, relative humidity, and pressure); pulse time, current, and voltage (for electrostatic discharge); mode, maximum acceleration, and excitation frequency (for vibration); and energy, dose, dose rate, and pulse width (for radiation).

9.2 NON-OPERATING RELIABILITY SIMULATION EXAMPLE

To illustrate the practical use of physics-of-failure concepts in non-operating reliability assessment, a multichip module example has been selected. The multichip module has a silicon substrate, two 3-D die stacks of ten silicon dies each, and TAB die-to-substrate inner interconnects. The dies are DRAMs, 140 x 300 mil^2 in size, operating at 1 MHz clock frequency, dissipating 0.22 W per operational die and 0.06 W per standby-operation die. Each of the dies in the 3-D stack has 21 peripheral leads at 16 mils pitch, TAB-bonded to a silicon substrate. The capacitors are 20 WVDC, 0.2 μF conductive bag capacitors with 20% tolerance, palladium-silver terminations, and an insulation resistance of 1,000 GΩ @ 25°C. The dies are attached to each other with an organic cement, and the die module is attached to the substrate with solder. Between the die stacks, two capacitors are attached to the silicon substrate with epoxy. The substrate has a silicon base, with an SiO_2 passivation. The metallization material is aluminum/titanium/tungsten; the dielectric material is polyimide (Dupont 2611). The substrate is housed in a Kovar case with a nickel undercoat and a gold overcoat. The substrate-to-lead interconnects (outer interconnects) are gold wirebonds, bonded to thin-film gold pads on the silicon substrate. The package has 60 Kovar leads with nickel undercoats and gold overcoats, which are glass-sealed to the case. The package has an electric-resistant seam-welded Kovar lid, again with a nickel undercoat and a gold overcoat.

Table 9.A, appended to the end of this chapter, provides highly correlative parameters such as environment, architecture, and material properties on which the failure mechanisms are dependent. References to failure mechanisms models are also given. The non-operating reliability analysis results translate into a conservative estimate of a non-operating life of more than 15 years.

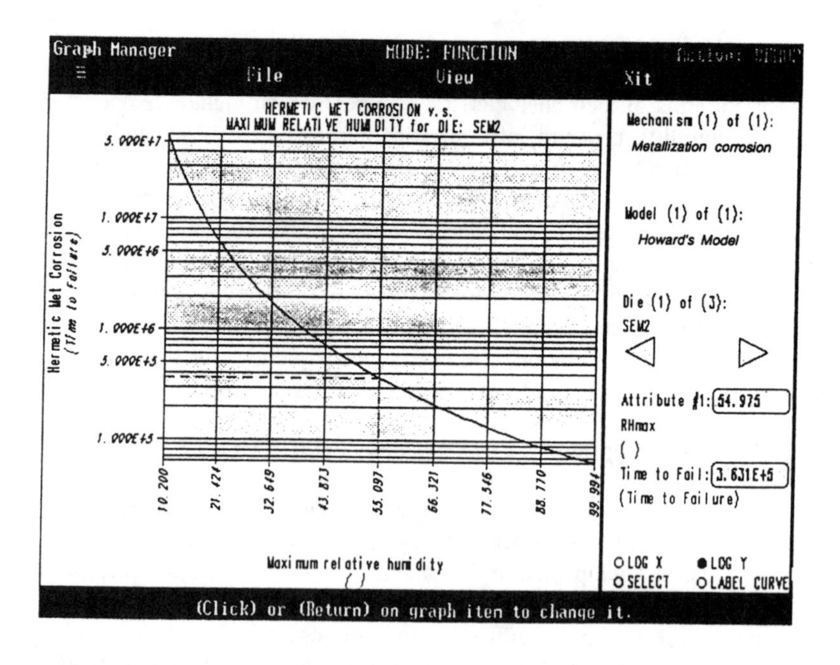

Figure 9.3 Sensitivity curves show the failure mechanisms changes resulting from attribute variations

Table 9.A Microelectronic Failure Sites, Failure Mechanisms, Key Stresses and Architecture Parameters, and References for Models.

Failure site	Failure mechanism	Highly correlative parameters during storage and dormancy		References for models
		Environment stress parameters	Architecture and property parameters	
Wire	Flexure fatigue	ΔT	$r_{\sim wire}$: wire radius $C_{f,D\sim wire}$: wire fatigue ductility coefficient $b_{d\sim wire}$: wire fatigue ductility exponent $\beta_{ang\sim wire}$: angle of wire with substrate $\alpha_{\sim wire}$: CTE of the wire $\alpha_{\sim subs}$: CTE of substrate $E_{\sim wire}$: modulus of elasticity of wire $l_{\sim wire}/d_{\sim wire}$: wire length/bond span	[Gaffeny, 1968], [Villela, 1970], [Ravi, 1972], [Phillips, 1974], [Pecht, 1989], [Harman, 1974]
Wirebond	Shear fatigue	ΔT	$E_{\sim bpad}$: modulus of elasticity of bond pad $E_{\sim subs}$: modulus of elasticity of substrate $G_{\sim bpad}$: shear modulus of bond pad $v_{\sim subs}$: poisson ratio of substrate $\alpha_{\sim wire}$: CTE of the wire $\alpha_{\sim subs}$: CTE of substrate $\alpha_{\sim bpad}$: CTE of bond pad $C_{f,D\sim wire}$: wire fatigue strength coefficient $b_{d\sim wire}$: wire fatigue strength exponent $C_{f,D\sim subs}$: wire fatigue strength coefficient $b_{d\sim subs}$: wire fatigue strength exponent $t_{\sim bpad}$: bond pad thickness $w_{\sim bpad}$: bond pad width	[Philosky, 1973], [Pecht, 1989], [Newsome, 1976], [Philosky, 1970, 1971], [Gerling, 1984], [Khan, 1986] [Pinnel, 1972], [Feinstein, 1979], [Pitt, 1982]

Failure site	Failure mechanism	Highly correlative parameters during storage and dormancy		References for models
		Environment stress parameters	Architecture and property parameters	
Wirebond	Kirkendall voiding	T, Contamination (Cl^-, Br^-, Na^+), t_i	Diffusivity of wirebond pad materials Rate constant, depending on the interdiffusion coefficients of the bonded materials Bond pad thickness Bonded wire thickness	[Newsome, 1976], [Philosky, 1970, 1971], [Gerling, 1984], [Khan, 1986] [Pinnel, 1972], [Feinstein, 1979], [Pitt, 1982], [Khan, 1986], [Villela, 1971]
Die	Fracture	Δ T, ∇ T	$E_{\sim subs}$: modulus of elasticity of substrate $E_{\sim att}$: modulus of elasticity of attachment $E_{\sim die}$: modulus of elasticity of die $\nu_{\sim die}$: Poisson ratio of die $\nu_{\sim att}$: Poisson ratio of attachment $\nu_{\sim subs}$: Poisson ratio of substrate $l_{\sim die}$: die length $t_{\sim die}$: die thickness $t_{\sim att}$: attachment thickness $t_{\sim subs}$: substrate thickness $\alpha_{\sim die}$: CTE of the die $\alpha_{\sim subs}$: CTE of substrate $a_{i\sim die}$: initial crack length in the die $a_{f\sim die}$: initial crack length in the die $A_{\sim paris\sim die}$: Paris' coefficient of the die $n_{\sim paris\sim die}$: Paris' exponent of the die $K_{C\sim die}$: fracture toughness of the die	[Tan, 1987], [Hawkins, 1987]

Failure site	Failure mechanism	Highly correlative parameters during storage and dormancy		References for models
		Environment stress parameters	Architecture and property parameters	
Encapsulant	Reversion	T	Reversion/ depolymerization temperature	[Tummala, 1989]
	Cracking	Δ T, dT/dt, RH	$l_{\sim die}$: length of the short side of the die $t_{\sim case}$: thickness of molding compound under the pad $P_{\sim case}$: vapor pressure in case cavity	[Nishimura, 1987], [Fukuzawa, 1985], [Kitano, 1988]
Package	Stress corrosion	dT/dt, RH	Residual stress, magnitude, and polarity of the corrosion galvanic potential Electrolyte concentration pH of electrolyte	[Tummala, 1989]
Device	Ionic contamination	T^{-1}, Contamination (Cl^-, Br^-, Na^+), ΔV_T: change in threshold voltage E: electric field	Q_m : mobile charge per unit area ϕ_F the difference in the Fermi level and the intrinsic Fermi level in the bulk of the semiconductor C_{ox}: oxide capacitance	[Brambilla, 1981], [Johnson, 1976], [Hemmert, 1980, 1981], [Bell 1980],

Failure site	Failure mechanism	Highly correlative parameters during storage and dormancy		References for models
		Environment stress parameters	Architecture and property parameters	
Die adhesive	Fatigue	ΔT	$E_{\sim subs}$: modulus of elasticity of substrate $E_{\sim att}$: modulus of elasticity of attachment $E_{\sim die}$: modulus of elasticity of die $\nu_{\sim die}$: Poisson ratio of die $\nu_{\sim att}$: Poisson ratio of attachment $\nu_{\sim subs}$: Poisson ratio of substrate $l_{\sim die}$: die length $t_{\sim die}$: die thickness $t_{\sim att}$: attachment thickness $t_{\sim subs}$: substrate thickness $\alpha_{\sim die}$: CTE of the die $\alpha_{\sim subs}$: CTE of substrate $a_{i\sim die}$: initial crack length in the die $a_{f\sim die}$: initial crack length in the die $A_{\sim paris\sim att}$: Paris' coefficient of attachment $n_{\sim paris\sim att}$: Paris' exponent of attachment $K_{C\sim die}$: fracture toughness of the die $b_{st\sim att}$: fatigue strength exponent $C_{f,st\sim att}$: fatigue strength coefficient	[Chiang, 1984], [Mahalingham, 1984]
Die metallization	Corrosion	dT/dt, T, RH	$w_{\sim met\sim die}$: die metallization width $t_{\sim met\sim die}$: die metallization thickness $R_{\sim sheet}$: sheet resistance of electrolyte $D_{e\sim met\sim die}$: metallization density	[Pecht, 1990], [Commizalli, 1980], [Inayoski, 1979], [Sim, 1979], [White, 1969], [Schnable, 1969]

Failure site	Failure mechanism	Highly correlative parameters during storage and dormancy		References for models
		Environment stress parameters	Architecture and property parameters	
Die metallization	Contact spiking	T	$D_{o\sim met}$: diffusivity of die metallization	[Chang, 1988], [Farahani, 1987], [T.I., 1987] [DeChairo, 1981], [Christou, 1980, 1982], [Ballamy, 1978],
	Constraint cavitation	T, dT/dt	$E_{\sim met}$: shear modulus of metallization $G_{\sim met}$: shear modulus of metallization $w_{\sim met}$: metallization width $D_{o\sim met}$: metallization diffusivity $T_{\sim pass}$: passivation deposition temperature	[Yost 1988, 1989]
Die	Electrical overstress	T	$\rho_{\sim die}$: die resistivity $D_{e\sim dop\sim die}$: doping density of die	[Alexander, 1978], [Canali, 1981], [Smith, 1978], [Runayan, 1965], [Pancholy, 1978]
	Electrostatic discharge	T I_p: peak current V_{device}: device voltage V_b: base voltage $V_{protection}$: voltage capacity of the device protection circuit	D_{base}: depth of base region $L_{emitter}$: length of the emitter region $T_{b\text{-}e}$: base-emitter separation C_b: body capacitance $R_{base\text{-}sheet}$: base sheet resistance R_i: arc resistance components τ: time constant of the RC circuit	[Kuo, 1983], [Hart, 1980], [Moss, 1982], [Amerasekera, 1986, 1987], [Scherier, 1978]

REFERENCES

Agarwala, B.N. Thermal Fatigue Damage in Pb-In Solder Interconnections. *23rd Annual Proceedings of International Reliability Physics Symposium* (1985), 198-205.

Ailles, C. R. and Neira, G. Dormant Storage Effects on Electronic Devices (ARFSD-CR-89014). U.S. Army Armament Research, Development, and Engineering Center, Picatinny Arsenal, New Jersey (August 1989).

Alexander, D. R. An Electrical Overstress Failure Model for Bipolar Semiconductor Components. *IEEE Transactions on Components, Hybrids, and Manufacturing Technology*, 1 (1978), 345-353.

Amerasekera, E. A. and Campbell, D. S. *Failure Mechanisms in Semiconductor Devices*, New York, John Wiley and Sons (1987), 12-96.

Amerasekera, E. A. and Campbell, D. S. Electrostatic Pulse Breakdown in NMOS Devices. *Quality and Reliability Engineering International,* 2 (1986), 107-116.

Arsenault, J. E. and Roberts, J. A. *Reliability and Maintainability of Electronic Systems,* Potomac, Maryland, Computer Science Press (1980).

Australian Ordnance Council Effects of Solar Radiation on Ammunition. Proceeding 14/83 (1983).

Attardo, M.J. and Rosenberg, R. Electromigration Damage in Aluminum Film Conductors. *Journal of Applied Physics*, 41 (1970) 2381.

Ballamy, W. C. and Kimmerling, L. C. Premature Failure in Pt-GaAs IMPATTs-Recombination Assisted Diffusion as a Failure Mechanism. *IEEE Transactions on Electronic Devices,* 25 (1978), 746-752.

Bell, J. J. Recovery Characteristics of Ionic Drift Induced Failures under Time/ Temperature Stress. *18th Annual Proceedings of the International Reliability Physics Symposium,* (1980) 217-219.

Blackford, P. A. and McPhilimy, H. S. Sand and Dust Considerations in the Design of Military Equipment, Report No. ETL-Tr-72-7, USAETL, Ft. Belvoir, Virginia (July 1972) AD-740135.

Blanks, H. S. Temperature Dependence of Component Failure Rate. *Microelectronics and Reliability,* 20 (1980), 219-246.

Brambilla, P., Fantini, F., Malberti, P., and Mattana, G. CMOS Reliability: A Useful Case History to Revise Extrapolation Effectiveness, Length and Slope of the Learning Curve. *Microelectronics and Reliability,* 21 (1981), 191-201.

Bratschun, W. R. and Wallner, J. L. Avoidance of Corrosion Problems in Consumer Electronic Products. *Electronic Packaging and Corrosion in Microelectronics*, (1987), 41-47.

British Telecom Handbook HRD3. Handbook of Reliability Data for Electronic Components Used in Telecommunications Systems, Issue 3 (January 1984).

Canali, C., Fantini, F., Zanoni, E., Giovannetti, A., and Brambilla, P. Failures Induced by Electromigration in ECL 100k Devices. *Microelectronics and Reliability,* 24, 1 (1984), 77-100.

Canali, C., Fatini, F., et al. Reliability Problems in TTL-LS Devices. *Microelectronics and Reliability,* 21 (1981), 637-651.

Chen, F. and Osteraas, A. J. Electrochemical Dendrite Formation During Corrosion of Connector Leads. *Electronic Packaging and Corrosion in Microelectronics,* (1987), 175-178.

Chen, K., Giles, G. and Scott, D. Electrostatic Discharge Protection for Micron CMOS Devices and Circuits. *In IEEE International Electron Devices Meeting Technical Digest,* (1986), 484-487.

Cherkasky, S. M. Long-term Storage and System Reliability. *IEEE Journal,* 3, No.1, (1970), 120-127.

Chiang, S. S. and Shukla, R. K. Failure Mechanism of Die Cracking Due to Imperfect Die Attachment. *Proceeding of the 34th Electronics Components Conference,* (1984), 195-202.

Christou, A., Report on the 1982 GaAs Device Workshop, *20th Annual Proceedings International Reliability Physics Symposium,* (1982), 276-277.

Christou, A., Cohen, E., and MacPherson, A. C. Failure Modes in GaAs Power FETs: Ohmic Contact Electromigration and Formation of Refractory Oxides, *19th International Reliability Physics Symposium*, (1981), 182-187.

Cieslak, W. R. Failure Analysis of 24-Pin Leaded Chip Carriers. *Electronic Packaging and Corrosion in Microelectronics,* (1987), 217-220.

CNET. Recueil de Donnees de Fiabilitie du CNET. (Collection of Reliability DATA from CNET), Centre National d'Etudes des Telecommunications (National Center for Telecommunication Studies) (1983).

Commizolli, R. B., White, L. K., et al. Corrosion of Aluminum IC Metallization with Defective Surface Passivation Layer. *18th Annual Proceedings of Reliability Physics Symposium,* (1980), 282.

Cottrell, D.F., et al. Dormant Operating and Storage Effects on Electronic Equipment and Part Reliability. *Technical Report No. RADC-TR-67-307,* (July 1967).

Cushing, M. et al. Design Reliability Evaluation of Competing Causes of

Failure in Support of Test-Time Compression. *Proceedings of the Institute of Environmental Sciences,* (1994).

Cushing, M., et al. Comparison of Electronics Reliability Assessment Approaches. *IEEE Transactions on Reliability,* 42 (December 1993), 600-607.

Davis, J. R. et al. *Metals Handbook: Corrosion*, 9th ed., 13, ASM International, Materials Park, Ohio (1987).

DeChairo, L. F., Electrothermomigration in NMOS LSI Devices, *19th Annual Proceedings International Reliability Physics Symposium,* 1981, 212-277.

DerMarderosian, A. Hermeticity and Moisture Ingress. Raytheon Company, Sudbury, Massachusetts, (1988).

Devanay, J. R. Failure Mechanisms in Active Devices. *Electronic Materials Handbook, ASTM International,* (1989), 1007.

Dummer, G. W. and Griffin, N. B. Galvanic Corrosion. *Environmental Testing Techniques for Electronics and Materials,* Macmillan Co., New York, (1962), 115-118.

Duvvury, C., Rountree, R., and White, L. A Summary of Most Effective Electrostatic Discharge Protection Circuits for MOS Memories and Their Observed Failure Models. *In Proceedings of the EOS/ESD Symposium,* (September 1983) 181-184.

Duvvury, C., Rountree, R., McPhee, R., Baglee, D., Hysop, A., and White, L. ESD Design Considerations for ULSI. *In Proceedings of the EOS/ESD Symposium,* (September 1985), 45-48.

Duvvury, C., Rountree, R., Baglee, D., and McPhee, R. ESD Protection Reliability in one Micron CMOS Technologies. *In Proceedings of the IEEE International Reliability Physics Symposium,* (1986), 199-205.

Duvvury, C. and Amerasekera, A. ESD: A Pervasive Reliability Concern for IC Technologies, *Proceedings of the IEEE,* 81(5), (May 1993), 690-702.

Edwards, D. G. Testing for MOS IC Failure Modes. *IEEE Transactions Reliability,* R-31, (1982), 9-17.

Engelmaier, W. Surface Mount Solder Joint Long-Term Reliability: Design, Testing, Prediction. *Soldering and Surface Mount Technology*, No. 1, (February 1989), 14-22.

Farahani, M. M., Turner T. E., and Barnes, J. J., Evaluation of Titanium as a Diffusion Barrier Between Aluminum and Silicon for 1.2 mm CMOS Integrated Circuits, *Journal Electrochemical Society,* 134, (1987), 2835-2845.

Feinstein, L. G. and Bindell, J. B. The Failure of Aged Cu-Au Thin Films by Kirkendall Porosity. *Thin Solid Films, 62,* (1979), 37-47.

Feinstein, L. G. and Pagano, R. J. Degradation of Thermocompression Bonds to Ti-Cu-Au and Ti-Cu by Thermal Aging. *29th Proceedings of the*

Electronic Components Conference, (1979), 346-54.

Flinn, P. and Chiang, C. X-Ray Diffraction Determination of the Effect of Various Passivations on Stress in Metal Films and Patterned Lines. *Journal of Applied Physics,* 67, (March 15 1990), 2927-2931.

Frankenthal, R. P. Corrosion in Microelectronics Current Status and Future Directions. *Electronic Packaging and Corrosion in Microelectronics,* (1987), 295-296.

Fukuzawa, I., Ishiguro, S., and Nanbu, S. Moisture Resistance Degradation of Plastic LSI's by Reflow Soldering. *Proceedings of the 23rd Annual International Reliability Physics Symposium,* (1985), 192-197.

Gaffeny, J. Internal Lead Fatigue Through Thermal Expansion in Semiconductor Devices. *IEEE Transactions on Electronic Devices,* ED-15, (1968), 617.

Gerling, W. Electrical and Physical Characterization of Gold-Ball Bonds on Aluminum Layers. *34th Proceedings of the IEEE Electronic Components Conference,* (1984), 13-20.

Hall, P. M., Panousis, N. T., and Menzel, P. R. Strength of Gold Plated Copper Leads on Thin Film Circuits Under Accelerated Aging. *IEEE Transactions on Parts, Hybrids, and Packaging,* PHP-11, 3, (1975), 202-5.

Harada, M., Tanigawa, S., Ohizumi, S., and Ikemura, K. X-ray Analysis of the Package Cracking During Reflow Soldering. *IEEE International Reliability Physics Symposium,* (1992), 182-189.

Harman, G. G. Metallurgical Failure Modes of Wire Bonds. *Proceedings of the 12th International Reliability Physics Symposium,* (1974), 131-141.

Harris, A.P. Reliability in the Dormant Condition. *Microelectronics and Reliability,* 20, (1980), 33-44.

Hart, A., Teng, T. T., et al. Reliability Influences from Electrical Overstress on LSI Devices. *18th Annual Proceedings of the International Reliability Physics Symposium* (1980), 190-96.

Hawkins, G., Berg, H., Mahalingam, M., Lewis, G., and Lofran, J. Measurement of Silicon Strength as Affected by Wafer Back Processing. *25th Annual Proceedings of the Reliability Physics,* (1987), 216-23.

Head, O. G. Drastic Losses of Conductivity in Antistatic Plastics. *EOS/ESD Symposium Proceedings,* (1982), 120-123.

Hirschberger, G. and Dantowitz, A. Evaluation of Environmental Profiles for Reliability Demonstration, Grumman Aerospace Corporation, Technical Report RADC-TR-75-252, RADC, Griffiss AFB, New York, (1975), AD-8007-946.

Howard, R. T. Environmentally Related Reliability in Microelectronic Packaging. *Electronic Packaging and Corrosion in Microelectronics,* (1987), 131-144.

Inayoski, H., Nishi, K., et al. Moisture Induced Aluminum Corrosion and Stress on the Chip in Plastic Encapsulated LSI's. *Proceedings of the 17th Reliability Physics Symposium,* (1979), 113-7.

Johnson, G. M. Accelerated Testing Highlights: CMOS Failure Modes. EAS-CON-76 Record, 142-A-142-I (1976).

Jowett, E. C. *Electrostatics in the Electronics Environment*, New York, John Wiley and Sons, (1976).

Jowett, C. E. *Electronics and Environments*, New York, John Wiley and Sons, (1973).

Ju, J. B. and Smyrl, W. H. Corrosion Studies of Thin Film Materials for Magnetic and Microelectronic Application. *Electronic Packaging and Corrosion in Microelectronics,* (1987), 119-125.

Khan, M. M. and Fatini, H. Gold-Aluminum Bond Failure Induced by Halogenated Additives in Epoxy Molding Compounds. *Proceedings of ISHM,* (1986), 420-8.

Kinsman, K. R. Integrated Circuit Packaging — A Materials Microcosm. *Electronic Packaging and Corrosion in Microelectronics,* (1987), 1-10.

Kitano, M., Nishimura, A., Kawai, S., and Nishi, K. Analysis of Package Cracking During Reflow Solder Process. *Proceedings of the 26th Annual International Reliability Physics Symposium,* (1988), 90-95.

Klinger, D. Nakada, Y., and Menendez, M. *AT&A Reliability Manual.* Van Nostrand Reinhold, New York, 1st edition, (1990).

Koyler, M. J. and Guttenplan, D. J. Corrosion and Contamination by Antistatic Additives in Plastic Films. *EOS/ESD Symposium Proceedings,* (1982), 99-102.

Kucera, V. and Mattson, E. Atmospheric Corrosion of Bimetallic Structures. *Atmospheric Corrosion*, New York, John Wiley and Sons, (1982), 183-192.

Kuo, W. and Kuo, Y. Facing the Headache of Early Failures: State of the Art Review of Burn-in Decisions. *Proceedings of the IEEE,* 71, (1983), 1257-66.

Kurotori, I.S. and Schafer, H.C. Summary of Selected Worldwide Temperatures in Explosive Hazard Magazines. *NWC TP-5174,* Naval Weapons Center, China Lake, California (February 1972), AD-892684L.

Lall, P., Pecht, M., Barker, D., and Dasgupta, A. Practical Approaches to Microelectronic Package Reliability Prediction Modeling. *Proceedings of the International Society for Hybrid Microelectronics Symposium*, Baltimore, (October 24-26 1989), 126-130.

Lall, P. and Pecht, M. An Integrated Physics-of-Failure Approach to Reliability Assessment. *Proceedings of the 1993 ASME International Electronic Packaging Conference*, EEP, 4-1, (1993), 509-524.

Livesay, B.R. The Reliability of Electronic Devices in Storage Environments. *Solid State Technology,* (October 1978), 63-68.

Lycoudes, N. and Childers, C. G. Semiconductor Instability Failure Mechanism Review, *IEEE Transactions Reliability,* 29, (1980), 237-247.

Mahalingham, M. et al. Thermal Effects of Die Bond Voids in Metals, Ceramic and Plastic Packages. *IEEE,* (1984).

Manson, S. S. and Hirschberg, M. H. *Fatigue, an Interdisciplinary Approach.* Syracuse, New York, Syracuse University Press, (1964), 133.

May, T.C. et al. Measurement of Alpha Particle Radioactivity in IC Device Packages. *Proceedings of the International Reliability Physics Symposium,* (1979), 13-21.

McAteer, J. O. *Electrostatic Discharge Control,* New York, McGraw Hill, (1989).

Melmed, M. Long-Term Inactive Storage Effects on Electronic Devices and Assemblies in Support of Sadarm. (ARFSD-TR-90015). U.S. Army Armament Research, (1991). Development, and Engineering Center, Picatinny Arsenal, New Jersey (November 1990).

MIL-HDBK-217. Reliability Prediction of Electronic Equipment, U.S. Department of Defense (1991).

MIL-STD-883C. Electrostatic Discharge Sensitivity Classification. Technical Report Notice 8, DOD (March 1989).

Minear, R. L. and Dodson, G. A. Effects of Electrostatic Discharge on Linear Bipolar Integrated Circuits. *15th Annual Proceedings of the Reliability Physics Symposium,* (1977), 138-143.

Moss, R. Y. Caution — Electrostatic Discharge at Work. *IEEE Transactions on Components, Hybrids and Manufacturing Technology*, 5, (1982), 512-515.

Neff, G.R. Hybrid Hermeticity and Failure Analysis. *Hybrid Circuit Technology*, 3, (1986), 19-24.

Newsome, J. L. and Oswald, R. G. Metallurgical Aspects of Aluminum Wire Bonds to Gold Metallization. *14th IRPS,* (1976), 63-74.

Nishimura, A., Tatemichi, A., Muira, H., and Sakamoto, T. Life Estimation for IC Plastic Packages Under Temperature Cycling Based on Fracture Mechanics. *IEEE Transactions on Components, Hybrids, Manufacturing Technology,* CHMT-12, (1987), 637-642.

Noon, D. W. Corrosion and Reliability Industrial Process Control Electronics. *Electronic Packaging and Corrosion in Microelectronics, (1987),* 49-53.

NTT Standard Reliability Table for Semiconductor Devices, Nippon Telegraph and Telephone Corporation, (March 1985).

Okabayashi, H. An Analytical Open-Circuit Model For Stress Driven Diffusive Voiding in Al Lines, Proceedings of the 5th International Conference Quality In Electronic Components; Failure Prevention, Detection and Analysis, (1991), 171-175.

Olesan, H. L. Radiation Effects on Electronic Systems. New York, Plenum Press, (1966), 35-87.

Pancholy, R. K. Gate Protection for CMOS/SOS. *15th Annual Proceedings of the Reliability Physics Symposium,* (1977), 132-137.

Panousis, N. and Hall, P. The Effects of Gold and Nickel Plating Thicknesses on the Strength and Reliability of Thermocompression-Bonded External Leads. *IEEE Electronic Components Conference,* (1976), 74-79.

Pecht, M., Lall, P., and Dasgupta, A. A Failure Prediction Model for Wire Bonds. *Proceedings of the 1989 International Symposium on Hybrid Microelectronics,* (1989), 607-13.

Pecht, M. and Ko, W. A Corrosion Rate Equation for Microelectronic Die Metallization. *International Journal for Hybrid Microelectronics,* 13(2), (1990).

Pecht, M. Handbook of Electronic Package Design. New York, Marcel Dekker, (1991).

Pecht, M. Integrated Circuit, Hybrid, and Multichip Module Package Design Guidelines-A Focus on Reliability, New York, John Wiley and Sons, (1994).

Pecht, M., Dasgupta, A., Evans, J, and Evans, J. Quality Conformance and Qualification of Microelectronic Packages and Interconnects, New York, John Wiley and Sons, (1994).

Peck, D. S. and Zierdt, C. H. Temperature-Humidity Acceleration of Metal Electrolysis Failure in Semiconductor Devices. *11th IRPS,* (1973); 24th IRPS, (1986), 146.

Phillips, W. E. Microelectronic Ultrasonic Bonding. G. G. Harman, Ed., National Bureau of Standards (U.S.), 400-2, (1974), 80-86.

Philosky, E. Design Limits When Using Gold Aluminum Bonds. *9th Annual Proceedings of the IEEE Reliability Physics Symposium*, Las Vegas, Nevada, (1971), 11-16.

Philosky, E. M. and Ravi, K. V. On Measuring the Mechanical Properties of Aluminum Metallization and their Relationship to Reliability Problems. *11th Annual Proceedings of the Reliability Physics Symposium,* (1973), 33-40.

Pinnel, M. R. and Bennett, J. E. Mass Diffusion in Polycrystalline Copper/ Electrodeposited Gold Planar Couples. *Metallurgical Transactions, 3,* (1972), 1989-1997.

Pitt, V. A. and Needles, C. R. S. Thermosonic Gold Wire Bonding to Copper Conductors. *IEEE Transactions on Components, Hybrids, and Manufacturing Technology*, CHMT-5, 4, (1982), 435-40.

Pompei, D. Climatological Conditions and Effects on Storage at Selected Conus Army Ammunition Depots (Report No. QAR-R-018) Picatinny Arsenal, New Jersey, (October 1985).

Ravi, K. V. and Philosky, E. M. Reliability Improvement of Wire Bonds Subjected to Fatigue Stresses. *10th Annual Proceedings of the IEEE Reliability Physics Symposium,* Las Vegas, Nevada, (1972), 143-149.

Renninger, R., Jon, M., Ling, D., Diep, T., and Welsher, T. A Field

Induced Charged-device Model Simulator. *In Proceedings of the EOS/ESD Symposium,* (September 1989), 59-71.

Resnick, M. and Riedinger, V. T. Munitions Testing at Proving Grounds and in Desert, Arctic, and Tropical Environments. *Technical Memorandum 1607*, Picatinny Arsenal, New Jersey, (1965).

Riordan, P. Weather Extremes Around the World, Report Number ETL-TR-74-5, USAETL, Ft. Belvoir, VA, (April 1974), AD-A000802.

Rooney, J. P. Storage Reliability, Proceedings of the Annual Reliability and Maintainability Symposium, (1989), 178-182.

Rosengarth, C. W. Corrosion Protection for Semiconductor Packaging. *Solid State Technology, 27,* (June 1984), 191-196.

Rozozendaal, A., Amerasekera, P., Bos, P., Baelde, W., Bontekoe, F., Kersten, P., Korma, E., Roomers, P., Kris, P., Weber, U., and Ashby, P. Standard ESD Testing of Integrated Circuits. *Proceedings of the EOS/ESD Symposium,* (1990), 119-130.

Runayan, W. R., *Silicon Semiconductor Technology*, New York, McGraw-Hill, 1965.

Scherier, L. A. Electrostatic Damage Susceptibility of Semiconductor Devices. *16th Annual Proceedings of the Reliability Physics Symposium,* (1978), 151-153.

Schnabble, G. L. and Keen, R. S. Aluminum Metallization — Advantages and Limitations for Integrated Circuit Applications. *Proceedings of IEEE,* 57, (1969), 1570.

Schnabble, G. L. Failure Mechanisms in Microelectronic Devices. *Microelectronics and Reliability,* 1, (1988), 25-87.

Schnable, G. L., Comizzoli, R. B., Kern, W., and White, L. K. A Survey of Corrosion Failure Mechanisms in Microelectronic Devices. *RCA Review,* 40(4), (1979), 416-445.

Sim, S. P. and Lawson, R. W. Influence of Plastic Encapsulants and Passivation Layers on Corrosion of Thin Al Films Subjected to Humidity Stress. *Proceedings of the 17th Annual Reliability Physics Symposium,* (1979), 103-112.

Sinclair, J. D. Corrosion of Electronics by Ionic Substances in the Environment. *Electronic Packaging and Corrosion in Microelectronics,* (1987), 145-154.

Smith, J. S., El Overstress Failure Analysis in Microcircuits, *16th Annual Proceedings International Reliability Physics Symposium,* (1978), 41-46.

Sparling, R. H. Corrosion Prevention/Deterioration Control in Electronic Components and Assemblies, U.S. Army Missile Command, Redstone Arsenal, AL, (1967).

Stojadinovic, N. D., Failure Physics of Integrated Circuits: A Review, *Microelectronics and Reliability,* 23, (1983), 609-707.

Suhir, E. Die Attachment Design and Its Influence on Thermal Stresses in the Die and the Attachment. *Proceedings of 37th Electronic Components Conference,* (1987), 508-517.

Suhir, E. and Poborets, B. Solder-glass Attachment in Cerdip/Cerquad Packages: Thermally Induced Stresses and Mechanical Reliability. *Transactions of the ASME*, 112, (1990), 204-209.

Suhir, E. and Lee, Y. Thermal, Mechanical, and Environmental Durability Design Methodologies, Electronic Materials Handbook: Package, Materials Park, Ohio, ASM International, 1, (1989), 45-75.

Tan, S. L., Chang, K. W., Hu, S. J., and Fu, K. K. S. Failure Analysis of Die Attachment on Static Random Access Memory (SRAM) Semiconductor Devices. *Journal of the Electronics Materials,* 16(1), (1987), 7-11.

Texas Instruments, Military Products: Products Spectrum Nomenclature and Cross Reference, *Designer's Reference Guide*, Dallas, TX, (1988), 4-10 to 4-11; 8-9; 12-29.

Tummala, R. R. and Rymaszewski, E. J. *Microelectronics Packaging Handbook*, New York, Van Nostrand and Reinhold, (1989).

U.S. Army Missile Research and Development Command. Storage Reliability of Missile Materiel Program, Storage Reliability Analysis Summary Report on Electrical and Electronic Devices, Redstone Arsenal, Alabama, 1, (1978).

Van Lint, V. A. J. et al. *Mechanisms of Radiation Effects in Electronic Materials*, New York, John Wiley and Sons, (1980).

Villela, F. and Nowakowski, M. F. Investigation of Fatigue Problems in 1-mil Diameter Thermocompression and Ultrasonic Bonding of Aluminum Wire. *NASA Technical Memorandum*, NASA TM-X-64566, (1970).

Villela, F. and Nowakowski, M. F. Thermal Excursion Can Cause Bond Problems. *9th Annual Proceedings of the IEEE Reliability Physics Symposium,* Las Vegas, Nevada, (1971), 172-177.

Wagner, L. C. Failure Analysis of Metallization Corrosion. *Electronic Packaging and Corrosion in Microelectronics,* (1987), 275-279.

White, M. L. Encapsulation of Integrated Circuits. *Proceedings of the IEEE,* 57, (1969), 1610.

Yost, F. G., Romig, A. D., and Bourcier, R. J. Stress Driven Diffusive Voiding of Aluminum Conductor Lines: a Model for Time Dependent Failure. Sandia National Laboratories, Report SAND 88-09 46, NTIS DE 9589-0010507 (1988).

Yost, F. G., Amos, D. E., and Romig, A. D. Stress-Driven Diffusive Voiding of the Aluminum Conductor Lines. *International Reliability Physics Symposium,* (1989), 193-201.

Zakraysek, L. Metallic Finish Systems for Microelectronic Components, *IEEE Transactions on Components, Hybrids and Manufacturing Technology*, CHMT-4, (1981), 462.

INDEX